being a mindful sceptic

Mark Dangerfield

Mindful Sceptics

Published by Mindful Sceptics

Springwood, NSW 2777, Australia

mindfulsceptics.info

AI Use Statement

This book was conceived, written and edited by a human who occasionally used AI tools to assist with grammar, editing, summaries and focused searches.

Contents

Preface

There is a niggle somewhere in your gut, a slight nausea, even a little anxiety. Maybe it is a sense of foreboding that something terrible is about to happen, or perhaps it already is, and you are unprepared. Despite all humanity's knowledge and technology, we are still making a mess of the planet, our only home. You don't have the tools to cope.

Your college courses were big on content and theory but somehow left you bewildered in the real world. Perhaps you are fed up with the half-truths, sensationalism and fake news designed to keep you addicted to the newsfeeds. Maybe you are over the socials with their trolls and idiots who don't have a clue but hog all the oxygen. But you can't just turn all that off because humanity is in peril. It seems negligent to be ill-informed.

Suppose you could shrug off all the doom-scrolling pain, the fear of the future, and your eco-anxiety over the world going to shit. What if you could free up some space, gain awareness and know instantly where to put your energy and how to improve things for yourself and the planet?

Well, this book outlines a tool that can help—a way of thinking to help you save yourself and the planet.

I want you to be a **mindful sceptic**.

A mindful sceptic is not a know-it-all who sounds like a Dick but is balanced, aware and assured of what is fact and fiction.

A mindful sceptic easily recognises a fallacious argument, however clever, and can choose to smile, walk away or take on the snake-oil peddlers and fake news talking heads.

A mindful sceptic is rarely duped or fooled and can avoid the embarrassment of believing in nonsense from the average John Doe but also knows when to take personal responsibility.

More important than all this is that a mindful sceptic knows the facts, what matters, and when it does.

But it doesn't stop there, for a mindful sceptic can take action and knows where to put their energy. And to cap it all, mindful sceptics have the essential skills to save humanity and the planet.

The good news is that in this book are some simple tools to implement these learnings. Be a mindful sceptic, and you will be calmer, more aware, and better able to make a real difference in your life. You could follow your curiosity and critical thinking to become the next world-famous sceptic who found ideas to help humanity overcome our global challenges.

This book is short, but it took a career to write. There are decades of backstory from my time as an academic ecologist at four universities, as an entrepreneur, and as an environmental consultant to businesses, bureaucracies and governments. I have been privileged to have lived in the UK, Zimbabwe, Botswana and

Australia and visited dozens more countries where I met many inspirational people. Academics, students, farmers, activists, civil servants and ministers had something to say and a lesson about scepticism. I cannot thank them enough.

For this book, I am endlessly grateful to Chris Scott, my collaborator at Mindful Sceptics, who shared his practical ecological knowledge, kind ear, and ability to keep us both sane. Charlie Wilson was a wise inspiration in the early days of these ideas. Special thanks to Kim Crawford, John Szymanski, and Milena Muscatello for reading and commenting on earlier drafts. Any errors of commission, omission, and chance are mine.

A writer must have people close for support and encouragement. I have them, and I love them dearly.

I dedicate this book to people I don't know. I greeted two of them briefly at a tourist park on the NSW coast, where they were staying in their large caravan. The air conditioner was blasting away, the windows were open, the television was on loud, and a large SUV was ready to pull the rig to the next destination.

They were not mindful sceptics, or they would have left their van and spent all the time they could on walks along the beach just a few hundred meters away from their pitch. I do this for them because it is not their fault.

Chapter One

Who is a mindful sceptic?

When Nelson Mandela was released from prison on Robben Island, I was in Gaborone working as an ecology professor at the University of Botswana. The campus was a few kilometres from the border with South Africa.

The Batswana were busy enjoying the fruits of their two huge diamond mines facilitated by a government in power since independence. At the same time, a repressive regime was about to transition to majority rule in the economic powerhouse next door.

They were giddy times.

My privileged position as an expatriate academic meant I lived in a suburban bungalow just north of the university campus. The dusty yard was tended by Chenziva, a young, not strictly legal migrant from Zimbabwe. In return for accommodation in the kaya, which was usually the maid's home from home, Chenziva would sweep, tidy, and do odd jobs.

Even though Chenziva was visibly working away at garden chores, every weekend, like clockwork, a young local, Keneilwe, or Kenny to his mates, would sidle up to the gate and call for the master.

"*Dumela boss, o tsogile jang,*" Kenny would announce himself loud enough for the whole house to hear.

"*Ke tsogile sentle, rra*. I am well," I would say as I ambled toward the gate, quickly exhausting my Setswana.

For the next few minutes, we would converse in impeccable English over the state of my truck and how essential it was to clean it. And Kenny would gladly exchange his labour to bring the vehicle up to speed for 20 Pula. I would thank Kenny for his offer but explain that Joseph—we couldn't call the migrant by his real name for obvious reasons—had already taken the bucket to the dust just yesterday.

At this point, Keneilwe would begin a long lament about how his mother was sick in Mogoditshane and he needed medication to take to her, or it was his grandmother's funeral tomorrow, and he had no money to make the journey to Mahalapye. Meanwhile, his aunt, who lived far away in Serowe, needed tin sheets to finish the roof of her breezeblock house before the rains. It was always this or some other tale of lack in his extended family.

No doubt there was a slither of truth in each of Keneilwe's requests, but his mother was never well, and his grandmother was buried many times during the months that the young man came calling. And in that time, his aunt must have built a mansion. In the early 1990s, 20 Pula was enough for a few days of hot food from the street vendors or a whole night of maize beer revelry, and I was a wealthy white man. Twenty bucks was not quite pocket change for me, but it was near enough.

At first, labour was exchanged for funds. I let Keneilwe into the yard and provided the cleaning utensils, and the hi-lux began to show off its metallic blue shine. But after a month or two, the weekly conversations had repeated their dance to a familiarity where he knew that I knew, and I knew that he knew. In the end, Kenny would arrive to call the master, I would come to the gate, and Kenny would hold out both hands to receive the 20 Pula, and we would both smile as he went on his way.

And I was a mindful sceptic.

Scepticism is the opposite of dogma. It is the ability to question ideas and think critically and objectively no matter what cultural, social or personal beliefs are in the mind.

Young Keneilwe was loose with his information to gain an advantage over me. My innate scepticism meant that I didn't take his statements as facts. I was wary of his claims. That I let him run with his partial truths and falsehoods is the essence of mindful scepticism.

That I conceded the possibility of argument over the facts to provide him with weekly sustenance became a choice.

That said, I was fully engaged with my inner sceptic. I had doubts. I was not prepared to take the information offered literally. There were questions in my mind, a particular uncertainty. In Daniel Kahnman's language, I would let my fast-thinking brain be held up, at least for a few seconds, by my slow-thinking brain enough to assess the facts or even see if they were facts.

In the case of Keneilwe, the truth was there in his demeanour, intonation, the cut of his jib, and some experience on my part of his ilk. Anyone travelling to tourist hotspots will be familiar with the local hustlers. As my slow-thinking brain assimilated the evidence before me and dipped into my experiences of similar instances, I could piece together the truth or something close to it.

Keneilwe needed the money. I had the money, and he had the skills to relieve me of it. My scepticism got me through the situation quickly, even if it took slow thinking.

And so we have a crucial attribute of the sceptic.

A sceptic is less likely to be duped, taken for a ride, pulled for a fast one, get fooled, or avoid embarrassment from believing in nonsense peddled by the average John Doe—or Keneilwe.

Sceptics can usually recognise all but the most clever, fallacious or fraudulent claim and can smile, walk away or take on the argument as the mood takes them. However, constant questioning can make the sceptic a bit of a Dick, even if the examination happens quietly in your head. Nobody wants to be that mean, aggravating, or otherwise disliked person.

And so we need the mindful modifier.

A mindful sceptic is not a know-all or a constant questioner but is balanced and assured of what is fact and fiction. She knows when to ask and when to keep her internal reasoning to herself. But before we delve into mindfulness, let's familiarise ourselves with the sceptic. Rather than formal definitions, let's begin with some quotes.

> *Briefly stated, a sceptic is one who is willing to question any claim to truth, asking for clarity in definition, consistency in logic, and adequacy of evidence. The use of scepticism is thus an essential part of objective scientific inquiry and the search for reliable knowledge.*
>
> Paul Kurtz, American scientific sceptic and secular humanist.

In responding to Keneilwe's hustle, I internally questioned the truth of his travel requirements and the quantity surveyor's report from his aunt's building works. I couldn't question the boy's logic, for the need was consistent with the request, but I had no evidence.

All there was was his word. This is not evidence.

Here is another quote to describe the sceptic.

> *Science is [...] a way of sceptically interrogating the universe with a fine understanding of human fallibility. If we are not able to ask sceptical questions, to interrogate those who tell us that something is true, to be sceptical of those in authority, then we're up for grabs*

for the next charlatan, political or religious, who comes ambling along.

Professor Carl Sagan, American astronomer and science communicator

Doubt, logic and evidence are powerful. The questioning sets up the doubt, the logic flow begins the enquiry, and the evidence provides the answers that lead to knowledge. But before reasoning happens, before the application of logic, there must be a liberal dose of questioning or, if you prefer, curiosity as to the truth.

A sceptic is one who prefers beliefs and conclusions that are reliable and valid to ones that are comforting or convenient, and therefore rigorously and openly applies the methods of science and reason to all empirical claims, especially their own. A sceptic provisionally proportions acceptance of any claim to valid logic and a fair and thorough assessment of available evidence, and studies the pitfalls of human reason and the mechanisms of deception so as to avoid being deceived by others or themselves. Scepticism values method over any particular conclusion.

Steven Novella, American clinical neurologist and Professor at Yale University School of Medicine

Scepticism is at the heart of the scientific method and is a critical foundation of scientific training. The scientist should learn the philosophy of science before the algebra, chemical equations, and laboratory techniques of the chosen discipline. Ingrained in that philosophy is the search for truth through reasoning. Scientists are trained sceptics.

Sceptics avoid being deceived. They use knowledge and skills from logic and understanding of evidence to arrive closer to the truth. They rely on this method rather than attaching to any specific conclusion. They tend to think slowly, which is a good thing.

As with all human biases, there is the risk of excess. Preoccupation with scepticism can mean endless questioning, which is tiresome. Fortunately, the sceptic uses a powerful skill that a mindful sceptic perfects—knowing when not to ask the evidence question. Sometimes, the truth is clear without the need for science and reason. Keneilwe was a hustler, it sparkled in his eyes. He was good at spinning the yarn with the heavy emotional tug on white man guilt.

He knew it and executed it perfectly on goodness knows how many unsuspecting expatriates. He probably had half a dozen every Saturday morning. I couldn't say how many 20 Pula notes passed into his outstretched palm, left hand politely held across his right. I found out later that at least one of my colleagues knew Kenny.

But some sceptics follow the three-year-old who has just found the infuriating logic of the "Why Mummy, why?" and questions everything. These sceptics risk not being able to attach themselves to any conclusion. When the evidence is patchy or weak, or the situation is complex, it can be hard to find consistency in the logic and decide when the evidence is meaningful. But an experienced sceptic also knows when to follow Friar Francis of Ockham and run with the simplest, most logical explanation without definitive proof.

> *Scepticism is a provisional approach to claims. It is the application of reason to any and all ideas—no sacred cows allowed. In other words, scepticism is a method, not a position.*
>
> The Sceptics Society

Scepticism is a method, not a position with no sacred cows. This definition is straightforward to read and very hard to implement in everyday life. We are surrounded by sacred cows, sheep and goats that belong to us, our partner, the neighbour, nonna or the prime minister.

Disassociating from this menagerie is part of the skill set of all mindful sceptics.

> *It turns out that the true meaning of the word scepticism has nothing to do with doubt, disbelief, or negativity. Scepticism is the process of applying reason and critical thinking to determine validity. It's the process of finding a supported conclusion, not the justification of a preconceived conclusion.*
>
> Brian Dunning, Executive Director, Skeptoid Media

The sceptic is not a peddler—someone other than the person who has decided on a position and sticks to it with super glue. Deniers and extremists are rarely true sceptics; they are dogmatists who make the true sceptic wary. Sceptics are questioners who know when to query and when to ask the right question. They also know how to apply logic to find the answer. They are critical thinkers engaged in the process of finding a supported conclusion.

If you prefer a formal definition, I use the one developed by Dr Steven Novella, host of *The Skeptics' Guide to the Universe* podcast and president of the New England Skeptical Society.

> *A sceptic is one who prefers beliefs and conclusions that are reliable and valid to ones that are comforting or convenient, and therefore rigorously and openly applies the methods of science and reason to all empirical claims, especially their own. A sceptic provisionally proportions acceptance of any claim to valid logic and a fair and thorough assessment of available evidence, and studies the pitfalls of human reason and the mechanisms of deception so as to avoid being deceived by others or themselves. Scepticism values method over any particular conclusion.*

Scepticism is not denial

Some people deny climate change, they don't believe that man put a human bootprint on the moon, and many other deniable facts and events. Denialists are not true sceptics; they are bogus.

> *...the bogus sceptic is, in reality, a disguised dogmatist, made all the more dangerous for his success in appropriating the mantle of the unbiased and open-minded inquirer.*
>
> Richard Cameron Wilson, New Statesman

The bogus sceptic runs with AIDS denial, Holocaust denial, climate change denial, and countless other claims that we collectively call post-truth and assume that by questioning these things, they magically disappear. This pseudo-sceptical behaviour has recognisable traits that we look into later. The key one, for now, is the cherry-picking of evidence that conforms to a pre-existing belief. And without the sceptic's skills to determine this selective use of facts—all tennis balls are yellow—it is hard to combat the denialist.

Tennis balls used to be white until fluorescent yellow became the default colour in the late 1960s because yellow was easier to see on the first colour televisions.

The reason for bogus scepticism is the opposite of the desires of the true sceptic.

"(bogus scepticism) centres not on an impartial search for the truth, but on the defence of a preconceived ideological position".

Richard Cameron Wilson, New Statesman

Unfortunately, the use of sceptic tools for denial by the unscrupulous has sullied the whole idea of being a sceptic. I can call you a sceptic as a slur. You are the negative troublemaker who denies everything. Scepticism has become a dirty word.

Denigrating scepticism is more than a shame; it is a loss of a mental technique we need everyone to have. Humanity must have sceptics to apply curiosity and critical thinking to interpret our ever-changing world and the precarious human predicament.

With so many acute social, environmental and economic challenges to tackle, humanity needs all its brainpower in tip-top shape.

The sceptic and the contrarian

Scepticism is the ability to question and run with your curiosity to understand the truth of matters. Yes, questions can sound in the first instance to be against something, but the idea is that you're looking for evidence that comes about through the application of logic.

Sceptics seek facts in favour or otherwise of the premise and argument; they look for clarity in definition, consistency in logic, and adequacy of evidence. The contrarian is against whatever threatens their opinion or worldview. Rather than being guided by curiosity, the contrarian is just against it. He likes to disagree with other people and express unpopular opinions.

Contrarians are challenging when they attach to a position that has no significant evidence in its favour and plenty of evidence to the contrary—the climate does not change is a topical one. When you question the non-existent logic of the contrarian, all they have is to double down, often with great fortitude and thick skin. That such tactics can elect a US president or give QAnon a platform attests to the tenacity of human willpower. A contrarian is a person who disagrees or is against. Look out for these people when they pretend to be sceptics.

A sceptic questions rather than disagrees. It may be that answers to the questions result in disagreement, and just asking the questions can come across as contrary, but the difference is in the outcome. A true sceptic is guided by the definition, logic and adequacy of the evidence that provides answers to the question. The prospect of agreement is subject to what information is available and how it is interpreted and put into context to come up with a logical conclusion.

The true sceptic begins with a neutral, open mind and is swayed or not by the evidence.

Sceptics can often be convinced by arguments they may initially question or oppose. They can be persuaded if there's sufficient evidence in favour of the argument or the proposition. Equally, they may be unconvinced and then say, '*Well, I need more information to believe what you're saying, so I won't just now. No hard feelings*'.

In other words, the sceptic might arrive at the contrary position. But the critical difference with the contrarian is that the sceptic is open-minded enough to accept that the argument presented may be valid. The contrarian enjoys any doubt and runs with it to their own ends. Evidence is central to the sceptic.

It's good to be sceptical

In a complex world full of shiny baubles that readily distract us from the truth, it is essential to be a sceptic. It has to be good.

Suppose I had responded to Keneilwe without scepticism. The first interaction would be acceptable. I would hear his tale of woe and hand over the 20 Pula, believing everything he said and responding kindly.

The second and third times of his asking would be similar, for the tales are new and the need obvious. His mother was sick, his family had to pay respects, and his aunt showed initiative in bettering her situation.

Then, the second funeral would come about. What? She had already passed. And didn't your aunt already put the roof on her rondavel? Shit. What a con artist this Keneilwe is; I should never have believed him.

This realisation is terrible news for Kenny. Next time he rocks up at the gate, he will likely get a sharp rebuke, maybe even some attention from the red setter. It's not good for me or Keneilwe or the dog.

A sceptic is less easily duped and can avoid the embarrassment of believing in nonsense from the average John Doe or the sharp hustling of Keneilwe. And this is just the start, the simple streetwise abilities that help with everyday interactions with others. Sceptics also benefit from being alert to similar fast ones from politicians, the media and used car salesmen. Being a sceptic is good because you will see people more clearly.

Should the sceptic look beyond the everyday interactions that make up modern living, there is also the view of the bigger picture. Sceptics are better equipped to see what is happening in the world. They see a different perspective than the one on the news feeds.

If my first thought when I saw a headline was to believe it, I would be in trouble. A closer inspection of any claims by following logic and reviewing the research evidence—more on how to do this later—is that many news headlines are misleading at best, or worse, lies. If my first thought is, "*How does the writer know this?*" I am in with a chance. If I begin by searching for **clarity, logic and evidence** in the argument, I am in much better shape to appreciate the words, even to decide if they are worth reading. Try it yourself next time you are sucked into doom-scrolling with your thumb.

To understand the life of a sceptic, we will frequently return to these and other big-picture benefits.

Before that, we need to explain the mindful qualifier.

The sceptic and the mindful sceptic

If the sceptic is less easily fooled or embarrassed by charlatans and continues to live an inquisitive life with ninja-level discernment skills, what more is there? Why mindfulness? In a word, pragmatism. Mindfulness makes all the questioning, logic, and evidence meaningful in the everyday.

Mindfulness is the secret sauce that adds three essential attributes to the sceptic's questioning mind and skill in evaluating evidence—awareness, values and action.

Awareness adds humanity to being a sceptic. Questioning Keneilwe's facts and concluding that he was pulling a fast one was not enough. Mindfulness helped me understand why he was trying it.

The sceptic in me decided quickly that there were no facts on offer, only his word. Whilst what he wanted was clear (clarity in definition) and why he wanted to trade his time for money made sense (consistency in logic), I had no way of discerning the truth of his claims beyond an innate judgement of character (inadequacy of evidence).

A sceptic could have stopped there and sent the youngster on his way on the back of too much doubt.

Enter mindfulness, **the art and skill of paying full attention to what is happening in and outside you, moment by moment, without judgment**.

Now, we are not just talking about the likely suspects from the new age section of the bookstore—ashrams, asanas, pranayama and days in a silent retreat on the mountaintop. Mindfulness for the sceptic is the essential moderator of the weakness in scepticism for constant questioning and linear cause-effect logic. Being aware of feelings inside and out makes scepticism so much more effective.

Mindful sceptics don't just ask questions; they know where to find the right question. They are aware of the situation, context and the relative importance of their scepticism to themselves and everyone else.

A mindful sceptic seeks the facts, what matters, and when it does... **without judgment.** I'll repeat that—what matters and when it matters.

This skill is crucial when applied to the modern human predicament of 8 billion people growing at 8,000+ per hour; all sucked into a vortex of economic growth, rampant energy use and inequality. Mindful sceptics become aware of this bigger picture and the many good, bad and ugly elements that make up the scene.

Hold on a minute.

Scented candles, Shakti mats, and hour-long yoga nidra practice are all very well, but humanity is built on judgment because everything we do is set against values. And you can't mindfulness away values. You have to embrace them.

Values are the powerful and deeply ingrained individual beliefs that act as driving forces behind our actions. They serve as unwavering guiding principles, shaping and influencing behaviour. And they might not be logical. We tend to embrace the values instilled in us as kids and use them to form the foundation of our identity. They are often closely tied to the culture they represent, and we gain confidence in them because we understand that they are not just personal choices but are integral to how society works, ours and everyone else's.

Adding mindfulness to the sceptic helps negate the risk of judgement that questioning can suggest, especially if the question goes against dogma, but it still allows for values. Knowing the inside and out means knowing we are creatures run on values. Embracing values widens the questioning.

Mindfulness makes it possible to see the bigger picture. Indeed, we can zoom out to the tapestry of global ecology and see it as made up of values and value tradeoffs where actions have consequences and then know how and when to take personal responsibility.

A mindful sceptic thinks about it, really thinks about it.

And a mindful sceptic knows that, even with quality evidence, values can get in the way of logic.

As necessary as questioning, discernment, awareness, and value recognition are to a mindful sceptic, this mindset has another crucial consequence. A mindful sceptic **can take action and knows where to put their energy**.

All the questioning, objectivity, and discernment set up choices—not just an array of options to select from but clarity about what the best choice will be in the moment and the future. Clarity is a powerful motivator for action.

Mindfulness adds the power of action to the sceptic.

Who is a mindful sceptic? Every quietly confident, engaged human searching for the truth.

Key points from this chapter

- Mindful scepticism is a valuable tool for navigating the modern world. It promotes critical thinking, intellectual humility, and evidence-based reasoning.

- Scepticism through curiosity and critical thinking is crucial to making well-informed decisions. It encourages questioning and scrutinising ideas, regardless of cultural, social, or personal beliefs.

- Mindful scepticism involves being open-minded, intellectually curious, aware of cognitive biases to balance scepticism, and receptive to new evidence and ideas.

- A mindful sceptic avoids overload from misinformation and fake news by fact-checking claims, relying on reputable sources, and ensuring the reliability of information.

- Mindful scepticism involves intellectual humility, recognising that beliefs are subject to change based on new evidence.

- A mindful sceptic is comfortable with uncertainty and complexity, avoiding simplistic answers and being willing to engage with diverse perspectives in public debates and political discourse.

- Mindful scepticism encourages engaging in constructive and respectful dialogue with others, even if their views differ, to refine one's beliefs and promote intellectual diversity.

- A mindful sceptic can take action and knows where to put their energy.

Chapter Two

A brief history of scepticism

The first sceptics

Imagine this for a moment. You are standing out in the open savanna in Africa. It's early morning on a spring day some 14,000 years ago. The musty smell of dust and elephant dung is on the light breeze. Around you are half a dozen of your tight-knit band that, even when everyone is together, numbers no more than 30 souls. The night was cold, but there was plenty of wood for the fire that the young ones were already replenishing because the fire must live all day. The baboons have been foraging away since first light with an eye for any insects, seeds or other edible morsels that their fingers can pick from the grass. You are a little envious of their low gait. It's easier for them to get at the tasty grubs. You are thirsty, and luckily, clean water is available from what will later be called the Okavango River.

As your eyes search for opportunities for food, what you see looks something like this...

You feel a little exposed out here on the pan, given that the lions roared nearby last night, but there is a good chance they moved off in the sun's direction. It pays to stay vigilant. Fortunately, you have your brothers and sisters and cousins with you, for there is safety in numbers in these parts. It makes sense if the baboons, the impala, and the zebra do it.

What you eat today depends entirely on how good you are at gathering and hunting with primitive tools—maybe a rock, although these are prized possessions in the sandy habitat. Perhaps you have a spear or bow with arrows tipped with acacia thorns and smeared with the poisonous insides of a beetle grub that lives in the shade of the tree with huge oval leaves. Even if you search far and wide, there is no supermarket, crops, goats, sheep, or chickens.

All very primitive and primal.

As you contemplate the scene, only one thing is on your mind. The dangers of snakes, spiders, scorpions, poisonous plants, bilharzia in the water, and a host of the more prominent likely suspects that could trample or bite you are nothing compared to the hunger in your belly. All you can think about is food. Getting sucked into a rosy, even romanticised vision of how our ancestors lived is easy. In reality, it was tough, very tough.

You set off into the bush with your comrades, hoping the elephants have not raided the marula tree with ripe fruits. With luck, an impala or, better, a kudu will be lame or sick and not yet run down by the hyenas. It might be the biggest feast for many days.

Meantime, there is the hunger that never goes away. Until it did. It took close to 14,000 years to reach modern times, but you and many of the 8 billion human souls now alive—some 7% of all the humans who have ever lived—now gather food for breakfast from places like this.

Modern humans, *Homo sapiens*, have come a long way in a very short time. We have learnt to feed ourselves and a global population in the billions. It is an unprecedented evolutionary phenomenon. No other organism has invented ways of gathering the production of nature and, with the assistance of exogenous fossil energy, channel most of it into food for a single species.

Gather food in small groups for 300,000 years, and then in roughly 10,000 years, learn to grow your own and bend the power of nature to ensure that humans can eat much of what she produces. No more bending down for grubs.

Just how successful this has become is staggering. One reliable estimate is that 96% of the land mammal biomass on earth today—the weight of the mammals we can see with our eyes—is made up of humans and their livestock, the animals humans manage for our benefit.

Here is what that looks like.

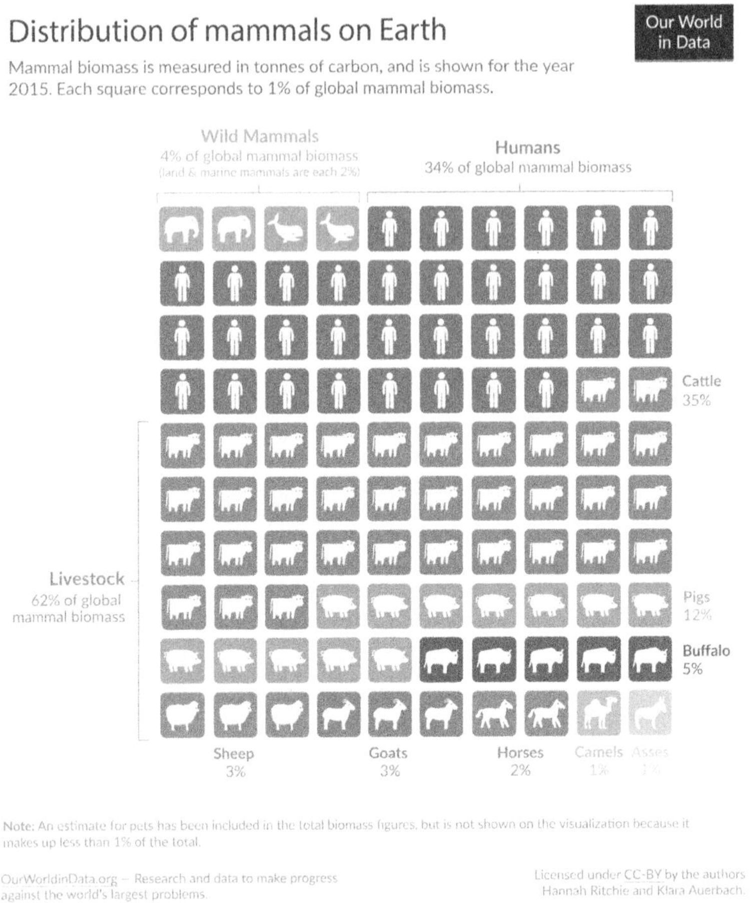

Livestock make up 62% of the world's mammal biomass; humans account for 34%; and wild mammals are just 4%. Data summarised from primary literature by Our World in Data.

It's unbelievable. But we are getting ahead of ourselves.

Before the advances in food production, our ever-hungry ancestors had managed to hunt and gather their way into almost every corner of the planet, even the cold or dry or windswept extremities.

Before the invention of agriculture, modern humans had already colonised the earth.

Whatever we were doing back then worked.

The explanations for the global success of *Homo sapiens* from social anthropologists, human ecologists, evolutionary biologists, and even the occasional philosopher will mention our biology—large brain, opposing thumbs, robust digestive physiology, vocal cords—and how these traits combined with the intelligence to control fire, use tools, cooperate and the ability to carry knowledge and skills across generations.

No doubt, all were crucial.

But let's go back to the savanna for a moment.

How did your little band know that the marula tree was in fruit? Or would the fruit not cause a horrible stomach ache? Who figured out how to attach a thorn to a pole and decided that such a weapon was useless unless the hollow tip contained a poison, enough to exhaust the kudu whilst it was still possible to track its anguished movements?

Who asked the questions that delivered productive answers?

The first sceptics.

Everyone is a sceptic at some time or another

We have all felt iffy about a claim to knowledge. Perhaps it was a cranky suggestion from a drinker in the pub, an old wives' tale, or a throwaway comment from a politician. There are plenty of folk around who will bend the truth a little to make a point, even your Nonna. Everyone has questioned conventional wisdom.

The original Greek meaning of *sceptikos* was 'an inquirer', someone unsatisfied and still looking for truth. In other words, the sceptic is a truth seeker and is not easily satisfied that the information presented or its interpretation represents the truth correctly. They want to be sure before they commit their belief to anything.

Sceptics raise the bar above the accepted truth to raise doubts about **knowledge beyond the contents of directly felt experience**.

This tendency to disbelief has been around for a long time. One ancient hunter was told by the head of his clan that only spears with acacia thorn tips were necessary to hunt antelope—this was the way.

Only the truth was it was hard enough to get close to a kudu to deliver the spear, and then the animal would run at the pinprick. The only chance of a kill was to deliver six spears simultaneously; even then, the animal was strong and could run far away. The truth was that *the way* wasn't very effective.

Curious, the ancient hunter questioned. There may be a better way. And there were several. But without the ability to raise doubts, the early hunters would not have figured out bows and poison-tipped arrows.

The truth is that we have yet to learn of what happened in prehistory when modern humans began their colonisation of the planet. Only snippets of fragmented evidence are available. Recorded history is different; it has written records. We know from texts that scepticism was present in ancient times when sceptics challenged the claims of Plato, Aristotle, and the Stoics. We also know that at the start of the Renaissance in the 1400s, when Leonardo Da Vinci painted the Mona Lisa and Michael Angelo crafted his statue of David, some sceptics questioned the philosophies of Scholasticism and Calvinism. Later, sceptics attacked Cartesianism and other theories, justifying a 'new science'. Later still, sceptics had a go at Kantianism and then against Hegelian philosophy. It was an endless round of questioning how the world worked.

What is clear from recorded history is that humans are sceptical creatures. We like to question the truth and have done it repeatedly, often against some heavy headwinds. Imagine how hard it would have been to think that the earth was round when all logic, practical experience and the bishop said it was flat.

As sceptics have appeared throughout recorded history, it is safe to assume that the tendency to question has been with us for much longer; perhaps it is innate.

The Age of Reason

Recorded evidence of scepticism goes back a long way. Still, it got big on us during the Enlightenment or Age of Reason, the period from 1715-1789, when European philosophers, thinkers and scientists built an intellectual and cultural movement that emphasised reason over superstition and science over blind faith. In this enlightenment period, scepticism meant disbelief—primarily religious disbelief.

Sceptics found enough confidence in the available evidence to tackle long-accepted fundamentals like the flat earth, the universe revolving around the planet, and creationism.

The Age of Reason also gave us the formal logic structure of the sceptic and the means to come up with reliable evidence for or against the conventions—**the scientific method**. This procedure takes the question and provides a means to discover a trustworthy, evidence-based answer. It is perhaps modernity's most potent tool, and we will talk much more about the scientific method in the next chapter.

For now, the arrival of the scientific method during the Enlightenment heralded a long line of notable and journeyman sceptics who questioned everything from the fundamentals of religion to the political and social order. This allowed scepticism to come in many flavours. Here are a few to give you the idea.

Epistemological scepticism

After the Age of Reason, "I told you so", or "Grandad always did it this way", was never enough. Thinkers went deeper. Some of them were so deep they nearly drowned and came up with questions like this: Can anything be truly known?

Only this is epistemology, the "*theory of knowledge, especially with regard to its methods, validity, and scope, and the distinction between justified belief and opinion.*" Before the scientific method and especially deductive reasoning took over, philosophers used epistemology—the theory of knowledge—to develop the strategies by which justified belief takes over from opinion.

For you and me, the ideas of knowing are the engine room of scepticism—where the propulsion comes from. However, for a time, the method of acquiring knowledge became a dominant form of scepticism—knowledge in general, questioning whether anything can be known with complete or adequate certainty—and was called epistemological scepticism.

Some sceptics were sceptical about how to be sceptical.

Academic scepticism

The brain has a huge computational capacity for abstract thought, including the ability to contemplate its death, which is raw power. Set this human mind free from everyday survival challenges, and you ask for trouble.

In the third century BCE, minds were allowed to roam free in the Platonic Academy, a school outside Athens founded by Plato in c. 387 BCE. Aristotle studied there for twenty years (367–347 BCE), and the Academy persisted for 300 years, way longer than most modern universities, before being destroyed by the Roman dictator Sulla in 86 BCE.

It was the Platonic Academy scholars who developed Socrates' observation, 'all I know is that I know nothing,' into the idea that nothing could be known. The philosopher who asserted that he knew some truth about the fundamental nature of things could not know any of his propositions with absolute certainty. The information we gain using our senses provides no guaranteed criterion or standard for determining which of our judgments is true or false.

The fundamental problem is that knowledge goes beyond the data. If the proposition might be false, then it would not deserve the name of knowledge but only that of opinion. Taking this premise to its logical conclusion, Academic sceptics said nothing is certain. The best information we can gain is probable and will be judged according to probabilities.

Today, if you follow the development of artificial intelligence (AI) with its neural networks and large language models, you will see what the Academic sceptics were about. Much of the success of current AI tools is that they embrace probabilities to deliver their answers. Sure, they train on vast information sets that the early Greek philosophers could have never imagined possible, but they judge according to probabilities.

I am not sure what is more remarkable. That ancient ideas reappear almost word for word as they were first imagined centuries ago, or that those toga-wearing students of the Platonic Academy figured it out. However, abstract thought was never going to stop at probabilities.

The Pyrrhonian school

If nothing can be known, then why bother asking a question? If something can be known, is there a problem with 'what something'? Either way, the thinker arrives at a point where no judgment is possible beyond what seemed to be the case. Mental anguish is inevitable.

One solution to this risk of insanity became known as Pyrrhonism, attributed to Aenesidemus (c. 100–40 BCE). Pyrrhonians proposed to suspend judgment on all questions on which there seemed to be conflicting evidence, including whether something could be known.

Scepticism for a Pyrrhonian was to suspend judgment on the question to avoid the risk of rashness and negative dogmatic conclusions. All you got was a sore arse from sitting on the fence.

The Pyrrhonian movement flourished up to about 200 CE, mainly in the medical community. People who thought they could know reality were constantly disturbed and frustrated by the next cadaver with its liver in the wrong place. However, by suspending judgment, they would neither affirm nor deny the possibility of knowledge but remain peaceful, still waiting to see what might develop.

Quite a few modern politicians are closet Pyrrhonians who can frequently neither confirm nor deny their position on a topic. Bit of a cake and eat it option, you would have to think.

Medieval scepticism

Religion crushed the Pyrrhonian movement, leaving the Christian Middle Ages to Academic scepticism—nothing could be known because only God knew the truth.

The view that truth in religion is ultimately based on faith rather than reasoning or evidence—known as fideism—appears in the late Middle Ages in the German cardinal and philosopher Nicolaus of Cusa's advocacy of learned ignorance as the way to religious knowledge. Most of the time, the power and awe of religion were enough to stifle the capacity for abstract thought. It was simply easier to believe in the divine.

Sceptical elements appeared in the followers of one of our ecclesiastical favourites, **William of Ockham** (1285–1347), in the fourteenth century—he of the razor and a prescient grasp of pragmatism. These people were brave enough to explore divine omnipotence and a divine source for all knowledge. Could God deceive humankind, regardless of the evidence, and could that make all human reasoning open to question?

True courage at the time.

Modern scepticism

Modern scepticism was in part thanks to William of Ockham—of two explanations that account for all the facts; the simpler one is more likely to be correct—but mainly from the rediscovery of classical scepticism, especially the Pyrrhonnian tradition to '*suspend judgment on all questions on which there seemed to be conflicting evidence, including the question whether or not something could be known*'. Be cautious, for you never know what is beyond the horizon. Even if God can see that far, her omnipotent vision is no help in the every day.

The voyages of exploration helped this caution along as the world became much bigger and more mysterious. Scholars also rediscovered the learning of ancient Greece, Rome, and Palestine, whilst many became proficient in the new thing called science from the Renaissance. The Renaissance and Reformation in the mid-1600s saw the publication of ancient texts on scepticism in Latin, the language of religion. Together, these developments combined to undermine confidence in the accepted picture of the world that soon became round and entered an orbit around the sun.

There was also a religious controversy between Protestants and Catholics around the basis of spiritual knowledge. Each was sceptical of the others' claims over doctrine—another delicious irony.

Historians record **Michel de Montaigne** (1533–1592) as the main revivalist for ancient scepticism, especially the Pyrrhonian arguments of complete doubt and their relevance to the religious debates of the time. I like Montaigne, not just because he was a critical philosopher of the French Renaissance but also for promoting the essay as a literary genre, especially merging anecdotes and autobiography with intellectual insight. The *Mindful Sceptics Guides* that support this book borrow heavily from this style.

Montaigne's scepticism was highly influential in the early seventeenth century, only to be questioned by other sceptics—this constant questioning can be a bit of a spiral. A contemporary of Montaigne, the English philosopher **Francis Bacon** (1561-1626), led the advancement of both natural philosophy and the scientific method to overcome the sceptical difficulties by appealing to new techniques and instruments that might correct errors and yield firm and unquestionable results.

As always with this history, the breakthroughs came from a version of the scientific method, the ability to create evidence that stood alone and did not need a value proposition or an opinion to provide it legitimacy.

Also, in this golden era of philosophy, **René Descartes** (1596-1650) applied the sceptical method of doubting all beliefs that could be false to discover truth—*cogito ergo sum*—I think, therefore I am. Consequently, actual knowledge is whatever is clearly and distinctly conceived is true. *Cogito ergo sum* was handy at the time because it meant that God's existence, that she is not a deceiver, that she guarantees our clear and distinct ideas, and that an external world can be known through mathematical physics. From scepticism, Descartes claimed to have found a new basis for certainty and knowledge of reality.

Ah, yes, but could Descartes' criterion determine what was true or false? Could we tell what was clear and distinct, or could we only suggest that something appeared clear and specific to us? Yes, there was criticism of Cartesianism, too. Proposal, criticism, and counter-proposal are what humans do, after all.

Various English philosophers, culminating in **John Locke** (1632-1704), in typically British fashion, tried to blunt scepticism with common sense given the 'reasonable' man's inability to doubt everything. They admitted that there might not be sufficient evidence to support knowledge beyond immediate experience, but this did not require that everything be questioned. All that was needed was a good dose of common sense.

Earlier, **Thomas Hobbes** (1588-1679) told everyone that the solution to the problem of finding the criterion for judging what was genuinely authentic was ultimately political—the sovereign would have to decide.

And we could go on. But it all got very messy once scholars began to present their thoughts in print, primarily via the mechanical printing presses that first appeared in the 15th century. Keeping up with philosophers' writing was challenging, and historians recorded scepticism being absorbed and merged into many fields of enquiry.

One outcome of all of the philosophising was that the consequences of all these philosophies influence all modern humans because they became the foundations, methods, and implications of science—specifically, what qualifies as science, scientific theories' reliability, and science's ultimate purpose. In other words, scepticism found a multi-tool.

Relevance of history to scepticism

What is the relevance of a way too brief history of known but rarely read philosophers for mindful scepticism?

I hope you can see that scepticism is essential in understanding the world's workings. It was critical to many breakthroughs and social advances that the philosophers gave a voice. And particularly in modern times when we are bombarded with half-truths and downright lies from leaders and people who should know better, it is invaluable.

What is clear from this history of scepticism is that questioning in the pursuit of truth has been the case throughout human history. Since we could write things down, some people have questioned common knowledge, opinions, and current concepts. And there is no reason to assume it began with the written word. I would guess that members of those ancient hunter-gatherer tribes were sceptics.

The fact that scepticism has gone on throughout recorded history is really important. It tells us how much scepticism has shaped our history and modern times. Questioning is vital to ensure that concepts with legs can stay the course and leave behind the spurious arguments and opinions that come unstuck. Ensure unhelpful ideas are steadily removed from our understanding, and the truth persists.

It also tells us that we can't live without scepticism. It is a critical part of how we run our societies, even when we think that scepticism is dead or is the breeding space for denial, or we're trying to give it a bad name.

Remember that the common reason that scepticism is derided is that closely held ideas and paradigms are under threat. The oil lobby, for example, knows that oil will not be our primary energy source in 50 years, perhaps even in 10 or 15 years, and so they're scrambling to ensure that they make as much money as possible before their time runs out. No doubt they will continue to claim that oil is a good source of energy and that it should still be used for as long as humanity exists—the truth is that oil is better as a source of raw materials than energy, but we will leave that for the *Mindful Sceptics Guides*. Either way, the oil men will resist any questioning of the status quo that is making them money. Anyone who has ever come up against a popular paradigm is in for a hard time.

Let's take you back to the African savanna, a long time before any of this epistemological mind-bending. Remember you are hungry and have limited headspace for anything other than the next meal.

Did you have room for scepticism? Probably not. However, there is no doubt that the development that went on in our hunter-gatherer times had much to do with people questioning what was happening at any given moment. It might not have been sophisticated questioning or deep thinking about where the universe came from and its relationships to God. Still, there was undoubtedly knowledge gained from the nagging of the sceptic. I dwell on this history because mindful scepticism is critical to our future.

If we don't understand scepticism and fail to apply it diligently and liberally across all aspects of our society, then we'll be the worst for it. Dogma will prevail, and no matter if that dogma causes pain and suffering.

The Influencer's New Clothes

In a bustling city, a famous social media influencer was known for her extravagant lifestyle, luxurious fashion choices and exceptional lips. She posted photos and videos showcasing her designer outfits, lavish vacations, and glamorous parties. People everywhere admired and envied her seemingly perfect life, showering her with likes, comments, and shares.

One day, a group of clever pranksters decided to take advantage of the influencer's vanity. They created a fake luxury brand and claimed to have designed the most exclusive and exquisite clothing line the world had ever seen. They photoshopped images of these imaginary clothes, describing intricate patterns and captivating details.

The influencer, always eager to be at the forefront of fashion trends, was immediately drawn to the new brand. She struck a deal to become the brand ambassador and proudly showcased the digitally manipulated images of the non-existent clothes on her social media accounts. Her followers, wanting to stay in the loop, praised the influencer's impeccable taste and eagerly discussed the intricate designs.

The influencer continued to post about the clothes, attending events and fashion shows while confidently claiming to be wearing these exclusive pieces. No one wanted to be the first to admit that they couldn't see the clothes, fearing they might be trolled or ghosted. Only when the social platform took down her posts for exposing a nipple did anyone ask questions.

A young designer couldn't hold back and finally commented on one of the posts, asking where they could purchase these incredible clothes. The influencer, caught off guard, tried to maintain the charade but struggled to provide concrete information.

The funny thing was that the influencer's credibility rocketed as more people began to question the reality of the brand with its see-through clothing. In the modern world, it does not matter that the new clothes were stitched in non-existent fabric, for followers are more concerned with following than fashion.

The influencer went on to set up an account on a less prudish platform and made millions from her fans.

Key points from this chapter

- Sceptical thinking has always been crucial in shaping society, even for the early humans, who relied on gathering and hunting for survival.
- The Age of Reason and the Enlightenment partly came about because scepticism challenged traditional beliefs and promoted reason and science over superstition.
- Various forms of scepticism contribute to developing critical thinking, questioning, human thought and the pursuit of knowledge.
- The scientific method is a powerful tool that produces evidence-based knowledge and understanding of the world.
- Mindful scepticism balances questioning ideas and paradigms to ensure progress, truth, and healthy growth for humanity's future.

Chapter Three

The Scientific Method

Scepticism is also about answers

Sceptics are driven by doubt and curiosity to ask many questions, which is the easy part. Anyone can question ideas and opinions. What makes sceptics unique is that they go on to think critically, seek evidence and evaluate it objectively, no matter their cultural, social or personal beliefs. They are masters of objectivity and intellectual humility.

Objectivity is easy to say because most of us like to think we are free of bias and opinion. We kid ourselves. Humans tend to have a narrow view of life because we live it in our own experience. That means we cannot be everywhere to see everything, so we struggle to see beyond what we have seen or heard.

We also have egos that need attention. We love praise and seek it for even the most trivial of traits. We dislike conflict unless it triggers our ego, and we tend to gravitate toward groups of people who look and feel like we do. After a time as a group member, we are likely to protect the group image relentlessly.

As youngsters, we are malleable, so our upbringing significantly affects who we are and how we live. And no matter what happens in that early development, we are given a set of values—we imbibe principles or standards of behaviour and build judgement over what is essential in life. However, our values are fluid and unique to each individual. Values define us, but we can also quickly go against them, especially after a night out on the town.

In other words, we are biased. We find objectivity very difficult.

We find questions easy, but the objective, critical thinking element of being a sceptic is up against our culture, family, and personal beliefs—and ego.

We are going to need help.

Thankfully, all the heavy thinking in the Age of Reason produced a tool that has proven super effective in generating reliable evidence of how the world works.

It's called the scientific method.

What is the scientific method?

The scientific method is **a standardised way of building knowledge about the natural world** and what we want to do in and with it—it is a vending machine for objectivity.

Questions are answered through a series of steps that are repeatable and carried out to generate evidence. The evidence is believable because of the logic of the process rather than the opinion of the person carrying it out. Not what I said but what I did to generate the evidence.

There are several versions of this 'standard way' with a huge literature that describes all the nuances. However, the most common and familiar scientific method consists of **systematic observation, measurement, experiment, and the formulation, testing, and modification of hypotheses**. These steps are sequenced and loop back to make the process iterative. This deductive reasoning version of the scientific method works to generate reliable evidence because the logic process is designed to locate cause and effect—it finds dependable and repeatable outcomes when A is added to B with or without the presence of C.

Then, with this outcome knowledge—what we now know to be the truth, at least within the experimental conditions—theories are formulated about how the world works. When these theories are tested repeatedly with further iterations of the scientific method and proven to hold universally, they become settled science and, in the broader ones, laws of nature.

Simple.

Here are the key steps in the scientific method and how they go together

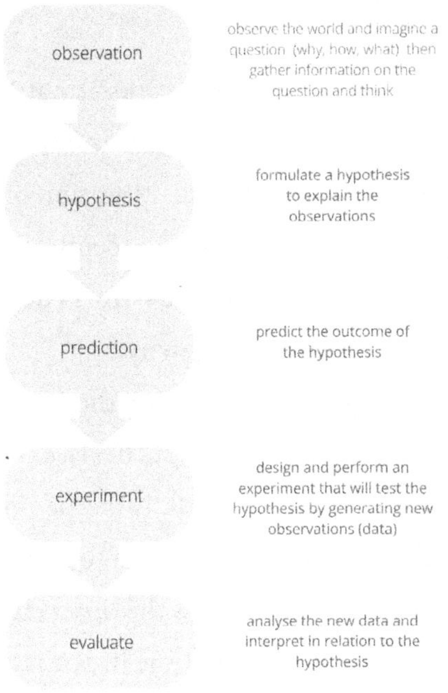

The details of each step can take a while to grasp. You can take a closer look at our *Mindful Sceptics Expert Guide to Understanding Evidence*, so here we will focus on their sequence and how they work together.

The deductive scientific method starts with an observation that triggers our curiosity.

Humans possess exceptional **observational skills**. We are especially good at spotting unusual elements in our surroundings. This ability evolved to help us find food and see the snake in the grass, whether that was the literal one or an unpleasant character in our tribe. Humans excel at capturing mental search images and actively seek more information about what intrigues us, especially the unusual.

Observation of the unusual sparks attention, but our curiosity will generate questions. Curiosity drives us to ask why certain things happen, leading to the identification of likely causes behind what we observe as unusual. For the sceptic, this can be the claim or opinion that sets off some doubt—**the crucial critical thinking is to turn doubt into asking why.**

The next step in the scientific method attempts to explain the observed phenomena with a **hypothesis**, a formal explanation for the observation. What we think is happening. Only for it to be scientific, the hypothesis must be testable. There must be a way to determine if the explanation fits the observation, which begins by knowing what might happen if the hypothesis were true.

Predictions are a crucial next step. They outline the expected outcomes and guide the selection of appropriate variables to manipulate the likely cause and measure the consequences. Usually, one prediction is for no change. Scientists call this the null hypothesis. In complex observations, multiple hypotheses and predictions can generate many consequences (effects) that require thorough measurements. Taking the time to make accurate predictions is always time well spent.

Next comes the test of the causal relationship implied in the hypothesis. In the standard (deductive) version of the scientific method, this is an **experiment** that involves systematically altering certain factors (independent variables that are expected to be the cause of the observation) while keeping others constant, observing the resulting changes (dependent variables that represent the observation), and collecting data to draw conclusions.

Experiments are many and varied and require experience to get right. They have known manipulation and experimental observations—recorded measurements of effects on control and treatment replicates. We will spare the details of controls, replication, randomisation, and the nuances of experimental design for *Mindful Sceptics Expert Guide to Understanding Evidence,* which you can access later. For now, it is enough to know that experiments can take various forms, from laboratory experiments conducted in a controlled setting, field experiments conducted in real-world environments to natural experiments that take advantage of unusual events in nature.

The critical next step is **evaluation**, which involves analysing the data from the experiment to determine any significant differences in observations between replicates in treatment and control groups. The key word is 'significant', which here asks the statistical question: Was the observation greater or less than would be expected by chance alone?

This is where everyone tunes out because we have entered the realms of probability and must employ statistics. Again, I will save the details and mental anguish for our expert guide. Still, whatever the statistical test, the idea is the same and straightforward enough. The statistical tests assess if the observed results in the treatment group result from the treatment itself or merely due to natural variation (chance) in the data. Comparison with control groups makes this analysis robust. Did the observations change significantly—to match the prediction or differ from the control—or not?

Evaluation allows the researcher to determine if the experiment has generated reliable evidence to test the experimental hypothesis. If the data aligns with the predictions based on the hypothesis, it is supported, and we accept the hypothesis. If the data do not align with the predictions, the hypothesis is not supported, and further investigation or a reformulated hypothesis is required.

In this way, the scientific method generates two feedback loops—one if the predictions are supported by the experimental evidence and another if they are not. This makes the scientific method iterative. It loops around to make evidence-gathering a process. Even if the hypothesis is supported, the scientists loop back to the prediction and test again. At least, that is what should happen.

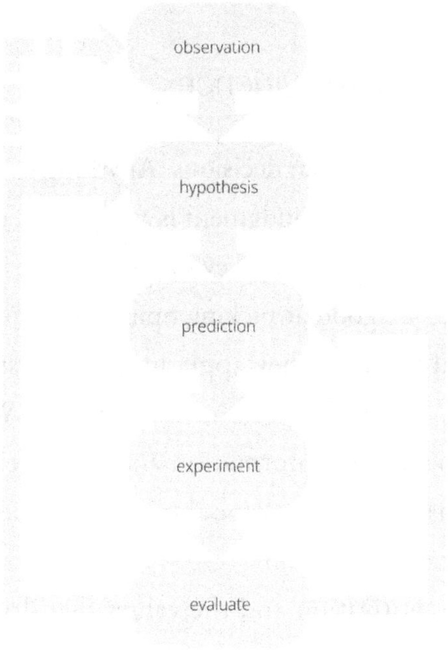

Whilst there are endless variations due to context, subject matter, and relevant skills, most scientists follow a procedure that is recognisable as observing, asking a question of the observation, formulating the question into a testable hypothesis, testing the hypothesis with an experiment, analysing the results to evaluate the hypothesis, rinse, and repeat.

As on Mandalor, this is the way.

Not all sceptics are scientists, and they don't need to be. Most of the time, sceptics are not looking to explain nature. They are asking questions about an opinion or a value judgment. They have to deal with values.

Science can evaluate evidence

Beliefs and values drive human decisions. Anything that contains a value proposition or a value judgment is an opinion, even if it is backed by detailed information and evidence.

Mindful sceptics are good at picking opinions because the curiosity and critical thinking they apply make the first question about motivation. Why is this person telling me this? What value or set of values makes this information and this interpretation important enough for them to tell me about it? Why did Keneilwe tell me his mother was sick, again?

Understanding motivation and the values that drive actions makes life much easier for the sceptic at risk of questioning everything. Knowing why Keneilwe was so forthcoming about his family's travails meant it was not necessary to determine the truth of his claims. More on values later, but the sceptic must apply doubt and curiosity to evidence.

What is presented as fact might not be. If a sceptic is told that the climate is warming, that the soil is degraded, or that eating more eggs will reduce the risk of heart disease, the first thought is not to believe each statement blindly but to ask, How do you know? What is the evidence for your assertion? How did you obtain the evidence? The first thought is not the conclusion—the soil is degraded—but what is your evidence for this assertion—soil carbon levels have declined by 50% in two decades—and how did you gather your evidence—a fully replicated, multi-factorial field experiment on the effects of cropping on soil carbon levels.

Whatever the philosophers did to confuse us all about the philosophies of science and theories of knowing, they all agreed on at least one thing.

There are different types of evidence.

Not all evidence is equal in strength, reliability and the confidence an observer can have in its efficacy to support or refute a conclusion—evidence is messy. Just as crucial to the scientific method as the evidence generation is how to evaluate it and deal with the scientific method's feedback loops. Sorting out the mess and deciding how helpful evidence will be to the argument is a challenge because all evidence from the scientific method arrives carrying a large bag labelled 'uncertainty'.

There are three broad sources of uncertainty: 1) Choice of the scientific method, 2) Application of the method, and 3) Interpretation of the results.

Let's take a brief look at each.

Choice of the scientific method

First, the scientific method is applied in various forms, each with a level of confidence in the outcomes. For example, controlled experiments are considered more reliable than observed comparisons, and evidence from a series of well-replicated, carefully designed experiments is more reliable than casual observations of the grass growing on either side of a fence or repeated measures of the carbon concentration in the atmosphere.

The method of science determines a large part of the evidence efficacy.

Application of the method

An experiment, an online survey, a field survey, a computer simulation, whatever the choice of the scientific method and approach to the actual work to gather the evidence that will appear in the form of recorded observations and measurements must have a **design**. The scientific method requires that the design is systematic, repeatable and consistent with the prediction. This applies to all branches of science and all levels of expertise.

Depending on the context, the design will cover attributes like sample size and representativeness, bias, confounding variables, and what is known about natural variability. And as with all designs, some are good for form and function, others not so much.

Then, the work gets done. There is a process of implementation that is a source of error. There are many opportunities for human mistakes, inexperience and variation between observers. There is also uncertainty from random or systematic errors and limitations in the techniques and instruments used to collect data and measure variables.

When poor designs are affected by application errors, the results can go haywire. The obvious implication is that error is part of doing science, and even the very best designs can be affected in execution.

Interpretation of results

After measurements are made, the numbers have to be understood. This data analysis and interpretation process considers the choice of the scientific method, the design, the many sources of error, and the expected level of inference.

The analysis turns data into information and allows the numbers to become evidence by evaluating against a likelihood of being true. We will see more about the likelihood in a moment. Data interpretation typically requires statistical testing according to the design and can include mathematical and statistical models. It is fun and involved, so there are good and plenty of not-so-good ways to do it.

That is not the end point once the results are in and interpreted. The evidence is used for **scientific inference**, the process of drawing conclusions or making predictions about the natural world based on evidence—the looping back from the evaluation step to the prediction. Scientific inference is essential for developing scientific theories, understanding causal relationships, and making evidence-based decisions. However, inference is not absolute certainty; it is a **probabilistic assessment based on the available evidence.**

And if the interpretation of results that suffer from uncertainty and the risks inherent in extrapolation and generalisation wasn't hard enough, there is the communication problem. A human has to undertake and tell us about the interpretation, which adds another layer of complexity. For example, publication bias occurs when studies with positive or statistically significant results are more likely to be published than those with negative or nonsignificant results. Again, several chapters in *Mindful Sceptics Expert Guide to Understanding Evidence* cover this topic of data interpretation.

All this is a bit much, so scientists have designed tools and frames to guide the sceptic towards the facts. There are several, and I am already going on a bit, but one crucial evaluation tool is worth mentioning here as it benefits mindful sceptics—the evidence hierarchy.

Evidence hierarchies

The evidence hierarchy in science refers to ranking or categorising evidence sources based on the strength, reliability, and confidence level in supporting a particular hypothesis, theory, or conclusion. There are many hierarchies in use by scientists, each designed to be more beneficial for certain disciplines or philosophies of science. Still, they all give categories for evidence from the lowest to the highest level of evidence reliability—the likelihood that the evidence offers rigour and how much it avoids bias.

Typical evidence hierarchies will include several of the following categories, starting with the most robust evidence source and down the list to the weaker inference:

- **Systematic reviews and meta-analyses** involve analyzing and interpreting data from multiple research sources and try to cover all that is known about a specific topic or question. These reviews used to be time-consuming and a logistical challenge, but modern search tools and databases make it possible for any sceptic to undertake one.

- **Randomized controlled trials (RCTs)** are considered the gold standard in scientific research, especially in medicine, because subjects are randomly assigned to receive different interventions or treatments, and outcomes are measured. Randomisation helps minimise bias, and the control group allows for comparisons and provides strong evidence for causal relationships.

- **Cohort studies** that follow a group of individuals over time to determine the development of specific outcomes are common in medicine and social sciences. They can identify risk factors and establish temporal relationships but are still observational and cannot prove causation definitively.

- **Case-control studies** are where researchers compare individuals (or subjects) with a particular condition or outcome (cases) to individuals without the condition (controls) and look for potential risk factors or exposures.

They can establish associations but not causation.

- **Cross-sectional studies** collect data from a specific population at a single time. They can identify associations between variables, but they cannot establish causation. Multivariate and factor analyses do similar things for environmental and social datasets.

- **Case studies** are anecdotal reports of individual cases and examples (or small groups of cases in medical research) that can provide interesting observations but only weak evidence due to their limited sample size and lack of control groups.

- **Expert opinion** is the lowest level of evidence hierarchies because while opinions or statements from experts in a particular field can be valuable, they are subjective and will vary based on individual expertise and bias.

Most scientists would advocate for sceptics and decision-makers to know about evidence hierarchies and to evaluate the quality and reliability of the evidence used to support critical thinking. The list above is typical for the medical sciences, where the evidence hierarchy was invented and most commonly employed.

I use a slightly different one for the ***Rapid Evidence Reviews*** that are foundational to our *Mindful Sceptics Guides* because most of the evidence comes from ecological and environmental research, where the research methods and application of the scientific method differs from that in medicine.

Mindful Sceptics
Evidence Hierarchy

1. Systematic reviews
2. Randomised trials and multi-site experiments
3. Natural experiments and time series of observation
4. Ecological process models and spatial models
5. Before and after studies, laboratory studies
6. Case studies
7. Expert opinion
8. Observations

Not just experiments

The version of the scientific method described here fits the Falsification Principle proposed by Austrian-British philosopher, academic and social commentator **Karl Popper** (1902—1994). Broadly, falsification suggests that for a theory to be considered scientific, it must be able to be tested and conceivably proven false. For example, the hypothesis that "all tennis balls are yellow" can be falsified by observing a white one from the day.

I hold a bias that my scientific training in the Department of Environmental Sciences at the University of East Anglia in the UK, which was rigorous, thorough and highly effective in training an ecological researcher, was steeped in this deductive reasoning. This wasn't surprising. My supervisors, and their supervisors, all came from the Oxbridge school of science thinking, where the Popperian philosophy that ***a theory in the empirical sciences can never be proven, but it can be falsified, meaning that it can (and should) be scrutinised with decisive experiments***, was the dominant paradigm. All my mentors did experiments.

One problem with deductive reasoning is that using experiments for falsification can never definitively refute a theory. There remains the possibility that the theory is correct and some other detail of the experiment was responsible for the negative outcome. The same problem troubled many philosophers we met in the last chapter.

A more tractable challenge is that whilst a well-planned and executed experiment is an excellent source of reliable evidence, not all observations can be tested this way. All global scale manipulations are impossible to implement, and no replicates exist. For example, we cannot experimentally test global climate change, only some mechanisms.

We know that the frequency and proportion of obesity is rising worldwide, and it correlates with the increase in grains, sugar, vegetable oils and processed foods in the diet, especially for those people with limited access to fresh fruit and vegetables. Any experimentation on the cause of this observation is torturous. Reliable experimental designs are immoral, and many confounding factors require massive replication.

And what about observations resulting from multiple interacting factors that occur throughout nature? It is challenging at best and often impossible to manipulate numerous causes simultaneously or in likely sequence to find the consequences of multiple causes. Randomised controlled trials might sit atop the evidence hierarchy in medicine, but experiments are not a panacea.

There are times when experiments are a problem.

Philosophies of science

Most textbooks on the scientific method focus on the falsification paradigm—deductive reasoning—that has dominated modern science. But it's not the only method of scientific enquiry. There are several other ways to gather evidence.

I like **inductive reasoning**, sometimes called inductivism. In this approach, the researcher collects specific observations, ideally lots of them, and through a study of the numbers that might include everything from simple correlation to complex multivariate statistical analyses, identifies patterns and generalises them into broader conclusions.

For example, regular, reliable, and repeated observation of atmospheric carbon dioxide shows a steady and steep increase in concentration over the last 100 years. Observations of temperature on land and in the oceans indicate that, on average, they are rising too. The correlation is strong, and the inductive implication is one causes the other—it helps that there is some experimental (deductive) evidence that carbon dioxide traps radiant heat.

Recent technologies have made inductive reasoning much more powerful. Computational capacity means we can analyse massive data sets quickly. Remote sensing from satellites and drones means we can make remarkable real-time and time-lapse observations of the planet to create interesting datasets for analysis. Artificial intelligence tools (AI) can help with the analysis in previously impossible ways. Combining these advances and analysing patterns can go deep and make inductive reasoning powerful.

Inductive reasoning is just one of several alternatives to the deductive version of the scientific method. What the sceptic needs to know is that there are several ways that science can help answer questions. Scientists and researchers may adopt different philosophical perspectives depending on the nature of their research and the questions they seek to answer. Evidence can come from many different philosophies of science.

The sceptic also needs to know that not all science is equal, and not all science follows the same precise approach to deduction. And that is crucial because the world is a complex, non-linear hotchpotch, and not all its secrets are revealed with an experiment.

Sometimes, it takes more than determinism to understand it.

Systems thinking

Back in the Renaissance period, when philosophers were throwing their scepticism around to find reason and challenge the church for the minds of the masses, the human enterprise was relatively simple.

When Leonardo Da Vinci painted the Mona Lisa in the early 1500s, the global human population was around 450 million. Fuel was charcoal and wood, power came from people and livestock, and the best way to get around was on a horse or a boat. There was limited international trade as the sailing ships were only just getting to far-off lands. Everything was primarily local. In this relatively simple world where humans were still mostly living within planetary means, you could go a long way on linear thinking.

Today, there are 1.4 billion people in India, another 1.4 billion in China and over 8 billion worldwide. The daily energy requirement from food is equivalent to 22 trillion kilocalories delivered across a six-continent supply chain. In 2022, global trade was roughly US$25 trillion, almost 400 times larger than in 1950, and incomprehensible to Leonardo. Moreover, economic activity is busting through many planetary boundaries.

Getting around is easier as there are roughly 1.45 billion vehicles on global roads and over 5,500 container ships in the global merchant fleet. Approximately 100 million barrels of oil are used—every day.

When a sceptic thinks about these numbers, a heap of questions appear. Her curiosity goes into overdrive, and she fires up the internet to help find evidence. But mindful sceptics also know that you can't experiment to the answers. Systems are complex, and human systems are replete with the unique psychology and sociology of people—they can be hard to understand.

Deductive reasoning is excellent, giving us a much deeper understanding of nature. It works exceptionally well for simple cause-and-effect scenarios and linear A+B=C type outcomes, so it helped humans become engineers and technologists. We wouldn't be here without it. Only cause-and-effect thinking is not a great way to look at systems. And it is systems that drive nature and the human enterprise.

When many elements interact with each other in complex, often unpredictable ways, falsification is challenging. It is hard to know what to predict, let alone put a number on it. Understanding systems—everything from food production to human health to koala conservation—requires analysis of the components of the system, their interconnectedness and their interactions. Instead of viewing individual elements in isolation, systems thinking looks at the entire system as a unified entity and explores the relationships and feedback loops among its parts.

7 Key principles of systems thinking

1. Interconnectedness—everything is interconnected, and changes in one part of the system can have far-reaching

effects on other parts.

2. Feedback Loops—reinforcing (positive feedback) or balancing (negative feedback) loops affect the system's stability and behaviour.

3. Emergence—new properties or behaviours emerge from the interactions of individual elements within the system and cannot be fully understood by studying the individual components alone.

4. Non-linearity—complex systems often exhibit non-linear behaviour, meaning small changes in one area can lead to significant consequences in other areas.

5. Holism—understanding comes from considering all the elements and relationships contributing to the system's behaviour.

6. Dynamic Complexity—the behaviour of a system can change over time due to various internal and external factors.

7. Boundary Definition—the system's boundaries are crucial in determining which elements are included in the system being studied and which are considered external factors.

A mindful sceptic is keen on evidence but knows that not all evidence comes from reduction to the delightful simplicity of cause and effect. And there is another problem that goes beyond complexity and the limitations of cause and effect logic—we have to return to the issue of values.

Science cannot decide

The scientific method generates information and evidence. It can help identify the cause of an effect and make comparisons possible. It has tools and procedures to measure attributes that humans recognise, but it does not create an opinion on those attributes—it should be neutral on values.

In a simplistic sense, science deals with true and false, whereas society deals with good and bad. Science can delineate the possibilities and describe the system likely to result from a policy, but it cannot decide on the morality or value of the resulting system.

> *Science can sometimes generate proxy measures of value but cannot make decisions based on value. That needs a human because values are beliefs, principles, or ideals that individuals or groups consider essential and desirable; they are subjective.*
>
> Lackey, R. T. (2007). Science, scientists, and policy advocacy. Conservation Biology, 21(1), 12-17

Scientists can objectively measure and provide information on attributes that humans find valuable. The scientific method can provide information about the consequences of actions or the effects of certain behaviours. But the evidence, even given the high-quality inference they provide, cannot make normative judgments about what is morally right or wrong, good or bad. **Neither science nor the scientific method can determine value in the ethical or moral sense.**

Values, ethics, and morality are subjective and cultural constructs that go beyond the realm of empirical observation and scientific inquiry. They involve human emotions, beliefs, and social norms, which are not easily quantifiable or testable. Different cultures and individuals may hold different values, and science alone cannot determine whose values are correct or superior.

For example, science can inform us about the environmental impacts of various energy sources. Extraction, distillation, and use of oil generate greenhouse gas emissions of 475 kilograms of carbon dioxide equivalent per barrel and deliver roughly 1,700 kilowatt-hours (kWh) of energy. Still, from a moral standpoint, it cannot be determined whether renewable energy sources with a fraction of emissions for the same power output are inherently "better" than non-renewable ones. The decision to prioritise one type of energy source over another involves value judgments extending beyond scientific investigation.

It is a choice.

Redrawn from data source: Carnegie Endowment for International Peace

Decisions on values are not true or false or even factual. Humans are known to decide on fiction as much as on fact.

Values are why we need science and the evidence it generates. It is there to help us decide, to let us know when we are pursuing a value and where we are choosing between true and false using deductive and inductive science at various levels of inference. Except that we are normative, we run on subjective values. That is the human condition. So science will always be overcome by emotion and the good and bad driving human decisions.

Science can't, but scepticism can help us understand when not to be consumed by these values and feelings, when not to be suckered and when not to run off with the fairies and latch onto the alternatives.

Values make scepticism essential

Scepticism is a combination of curiosity and critical thinking that finds objectivity. Asking if scepticism is good or bad is the wrong question. A better question is, do we need scepticism?

Yes, absolutely.

Every day, humanity faces bigger, more complex, more wicked challenges than ever. We are perilously close to self-destruction, and we feel it as we go about our business. Given this anxiety, some people will put their faith in a wing and a prayer—hoping for success despite a lack of preparation—while others will deny the reality because it cannot be accurate. Some will run away into the hills and form a commune that starts well but ends badly.

Many hardcore believers in the human enterprise will continue business as usual because it has gotten us this far, which is a hell of a long way, singing, 'Don't worry, the market will fix it.' The majority will likely follow this path because they have little choice. But if you ask Keneilwe if the market makes a living for him, he will laugh.

Here is the thing.

Change is upon us. It is happening all around because of what humans have done in the last 200 years. It's not easy to grasp the magnitude of energy and resource use by 8 billion humans, increasing at over 8,000 per hour, but the truth is right there in the device you are using to view these words—manufacture of the technology marvel that delivers an ebook to your retinas needed at least 70 of the 83 stable (nonradioactive) elements and 60% of the elements in the periodic table.

The sceptic has a different response to the world than most people. She doesn't run, hide, ignore or carry on regardless. She asks why. She follows her curiosity and backs it with critical thinking to find and evaluate the evidence. She learns the facts from the fiction and the evidence from the values. She is aware and empowered.

And here is why scepticism is essential.

She can take this power and apply it to help find solutions to human predicaments.

Key points from this chapter

- The deductive version of the scientific method is the dominant, standardised way of building knowledge about the natural world, and it involves systematic observation, measurement, and experimentation to generate reliable evidence.

- Scientific evidence has sources of uncertainty.

- Scientific inference is the probabilistic assessment of evidence, and it is essential for drawing conclusions about the natural and human world.

- The evidence hierarchy in science ranks different types of evidence based on their strength, reliability, and confidence in supporting a particular hypothesis or conclusion.

- Philosophies of science, such as falsificationism, inductivism, empiricism, rationalism, positivism, and more, provide different approaches to acquiring scientific knowledge.

- While the scientific method and deductive reasoning are valuable tools for understanding cause-and-effect scenarios, they are less effective for analysing complex systems with multiple interacting elements.

- Systems thinking is essential for understanding complex systems involving interconnectedness, feedback loops, emergence, non-linearity, holism, dynamic complexity, and boundary definition.

- A mindful sceptic understands the limitations of reductionist thinking and the complexities of human systems, leading to a well-rounded approach to seeking answers.

- Mindful sceptics add awareness to objectivity, which is crucial for evidence-based decisions over values.

Chapter Four

Misuse of Evidence

Is scepticism good or bad?

I am a highly trained and experienced sceptic. I know, it's a problem.

At college, I studied Environmental Sciences at a university that was so progressive one of the courses I took required that I grade my own assignments. After one of the more significant reports about constructing a management plan for a local nature reserve, the professor called me in and suggested I might have been harsh on myself. The work could have been better, but it merited a higher grade than I had given it. He told me that most students should have marked their work higher. Only a handful inflated their prowess. Odd that.

I enjoyed the freedom to learn so much that after my undergraduate program, I completed a research degree in ecology and delighted in an academic career where, as a college professor, I could let the students run with their ideas. I also completed and published ecological research, much of it on the ecology of soil animals, and climbed the academic tree.

It was a hoot.

My innate scepticism and passion for critical thinking did mean I ended up with some of the trickier teaching tasks. I taught a biostatistics course that contained more about the scientific method than probabilities. My ecology and biodiversity classes were replete with questions and open-ended learning. I always talked about objectivity, logical consistency and the importance of evidence.

True to my progressive roots, I even ran a practical class where the students were given sampling equipment, access to a 5 ha parcel of eucalyptus woodland and one question—How much biodiversity is there here? That was the extent of the instruction. Each year, the students came up with their version of an all-taxa biodiversity inventory, and we just had to write about it.

> *Dangerfield, J. M., & Pik, A. J. (1999).* The educational value of an all taxa biodiversity inventory. *Journal of Biological Education, 33(2), 76-83.*

Just as all this was swimming along, I gave it up.

Terrified by the thought of another 30 years locked up in the ivory tower, I shifted to an entrepreneurial career. Bizarrely, I wanted to build a tech company to tackle the sacred cow of environmental monitoring.

The assumption was that better environmental decisions could be made if the decision-makers had more reliable evidence. If the consequences of funding and land management decisions were more evident, they might even ask better questions and become more sceptical— today, this approach is called evidence-based policy.

This was less of a hoot.

There wasn't any money in ecological monitoring—there rarely is in actions people are forced to do—and the company came to an end, squeezed out by cash flow. This forced a second pivot to science consulting, where I had to sell my expertise instead of environmental evidence to commercial, government and education clients.

The hoot became a whimper when the emerging carbon markets hijacked this expertise, and I ended up advising on carbon offset projects in forestry, agriculture and land use. We all have to follow the cash when there are bills to pay.

My brief bio shows that even for the experienced specialist, asking questions and objective thinking is a hard gig in the real world.

So here is a thing about scepticism.

When used thoughtfully and critically, scepticism is beneficial and an essential approach to help understand the world and make informed decisions. It encourages individuals to question information, beliefs, and claims to promote a healthy sense of doubt and curiosity. Scepticism can help prevent us from falling victim to misinformation, scams, or false beliefs, leading to more rational and evidence-based choices.

This 'good' scepticism implies that people want to be thoughtful, informed, and have a healthy sense of doubt. These are risky assumptions.

Then, there is another thing about scepticism. Scepticism can have negative consequences when taken to an extreme. Excessive questioning can lead to a general distrust of all information, including credible sources and experts. This can hinder progress, create unnecessary doubt in well-established facts, and prevent individuals from accepting evidence-backed solutions. We will see later how this leads to the misuse of evidence and scepticism's terrible reputation in some sections of society.

The truth is that when scepticism is balanced and applied appropriately, it is a valuable tool for critical thinking and decision-making. It fosters a healthy questioning of information and encourages a search for evidence and truth. However, the rigid sceptic can dismiss well-supported facts and expert knowledge. Constant questioning can be irritating, and there are no answers and limited evidence for many of society's big questions. Sometimes, this is because they are questions about values and sometimes because there is no easy way to come up with evidence. Asking questions when there is no answer and no way to find a solution is asking for trouble.

In these situations, excessive scepticism is terrible.

So, is scepticism good or bad?

It is both. Curiosity and critical thinking are essential, but excessive questioning frustrates people almost as much as a lack of evidence. Finding the right balance is crucial for using scepticism to its full potential.

And because it is both, the sceptic can have a hard time. It takes courage to be a sceptic, and yet it is essential.

Scepticism is unavoidable

People like what they know because they are comfortable in their tribe and everyday life routines, regardless of their difficulty. Change or challenge is uncomfortable. We imagine that everything can stay the same, just as we like it. And yet, nothing is as we want it, and it only takes a little honesty to acknowledge that change always happens.

So, where does this contradiction come from?

Recall the African savanna that we asked you to imagine standing in with your fellow tribe members on a clear morning with the scent of elephants in the breeze. Nothing much changed if you were born into that tight-knit group of hunter-gatherers and survived to live an expected life of maybe thirty-odd years. The savanna was the same when you were born as when you passed on. The weather was just as always; the occasional wildfire had been through the grassland, the flood that filled the plains between the acacia islands arrived in late winter like it always did, and there was forever a snake in the grass. Life was demanding but predictable, stable and livable.

What you experienced was the same as that of your parents, their parents, and dozens of generations before. Nothing changed, just as we like it. And nothing changed for thousands of years after your brief time foraging for insects and dreaming of feasting on a kudu.

Until it did.

Roughly 12,000 years ago, during an especially warm climate phase, a few bright human sceptics in the fertile crescent of what is now the Middle East, over 8,000 km to the north of your savanna, figured out how to cultivate grain crops. Other bright humans in China and South America did a similar thing. This invention changed the lives of the communities where crops were grown. They invented different cultural norms, governance, and means to beat the crap out of their neighbours. This was critical as crops and the permanent buildings constructed beside them were worth protecting.

The change happened because the invention of agriculture created an energy surplus—more energy came out (food) than energy put in (human and later animal power). The surplus was a massive breakthrough because humanity broke our innate ecology's energetic shackles. And being an intelligent, organised ape species, we figured out how to use that energy surplus—we made more people.

But for a long time, this answer to the question of how to live differently and in one place did little to change the lives of most humans. Your descendants on the savanna in southern Africa had yet to learn what was happening elsewhere. Nor would they for several thousand years.

By the time they found out, all sorts of other changes had happened, from the domestication of livestock to the wheel, sailing ships and people paid to be soldiers. And, of course, the most significant change of all—the discovery of a vast new source of exogenous energy.

But we are getting ahead of ourselves.

During the hundreds of thousands of years that modern humans were few, persisting among several other species of Hominid and steadily figuring out how to make a living in all sorts of habitats, we only needed the occasional sceptic. Foraging and the hunt worked well enough.

We must have been curious to find new things, a better club or a more potent poison, and to travel. We were innate wanderers, and enough of us had the move-away gene to see us spread across the planet. Sceptics who didn't believe you could bring down a mastodon with a poison-tipped arrow were annoying but valuable tribe members. Still, for the most part, the predictability of the habitats and cultural norms for how to persist in them was enough.

Once humans settled in one place, many new problems emerged that couldn't be fixed with the old knowledge. Curiosity and critical thinking became valuable to finding solutions to the unique challenges of staying put, including a year-round supply of water, food, shelter, and new forms of defence.

Agriculture did it for the food and the rivers and lakes for the water, but new ways of doing things and new skills for living together permanently in the same place were needed. Many of the old ways had to be challenged.

Scepticism over how things used to be done was essential and unavoidable if the newly formed agricultural societies were to persist. And they did, adding more and more sophistication as time passed, much of it thanks to sceptics' awkward questions and critical thinking.

As humans gained what we call progress, more and more things changed. Now, it was essential to be sceptical as the foundation for knowing what worked and what didn't. The many philosophers throughout history scratching their heads and arguing amongst themselves were crucial to progress. They delighted in cogitation, and society had the time to absorb what they said.

Then, fossil fuels put a rocket under progress.

In a matter of decades, change accelerated to a point where it was hard for the most ardent sceptics to keep up. Humanity entered the industrial age and, in a blink the information age on the back of ubiquitous, cheap fossil energy. Along with it came capitalism, a system of organisation so vastly different from those that bound us together on the savannas that almost everything changed. A whole new set of challenges appeared.

Here is one of them.

Spinning Around

Paul Joseph Goebbels was a German Nazi politician and Reich Minister of Propaganda of Nazi Germany from 1933 to 1945. We are talking about one of Adolf Hitler's most devoted associates with skills in public speaking who was a vicious antisemite and widely held as responsible in large part for the attempted extermination of the Jews in the Holocaust. All this evil we remember.

But think about his substantive position for a moment—the Nazi party had a Minister for Propaganda.

Propaganda is *information, especially of a biased or misleading nature, used to promote a political cause or point of view.* In other words, a form of communication to distribute information that is always biased and designed to make people feel a certain way or believe a particular thing. Misleading the people is what, in hindsight, we would expect from an authoritarian regime needing to control a populous wearying of war and impending defeat. The problem is that we get something similar in all modern public relations; only these days, we call it spin.

> *spin is a form of propaganda, achieved through knowingly providing a biased interpretation of an event or campaigning to influence public opinion about some organization or public figure... "spin" often implies using disingenuous, deceptive, and manipulative tactics.*

And spin is everywhere: in marketing, in the media, on our socials, and especially in politics. If you own a smartphone, you cannot escape it.

Nowadays, a career as a Minister for Propaganda is a little off, even for post-truth politics. Still, you can have a career as a spin doctor, the guy or gal who **uses the facts and the evidence**

loosely and selectively enough to reduce negative perceptions and promote the agenda of whoever hires them.

In short, lies to make us feel something we otherwise wouldn't.

Spin is often delivered through a catchy phrase, slogan or sound bite that emphasises the required agenda or how people are expected to feel—Make America Great Again; Get Brexit done; Just stop oil; Save the whale.

Spin tactics often include selective use of the truth, particular quotes from speeches and press releases, burying negative news or sending it out on the Friday night before a public holiday.

It's an art form that does what all art tries to do—reach human emotions and mess with them. And unfortunately, the best modern spin doctors are true artists with a set of brushes and a wide colour palette. The sceptic must be wary.

Here are some of the tactics to look out for.

Economical with the truth

One way to spin is to use selective disclosure and be '*economical with the truth*'. This phrase goes back hundreds of years but was famously used by Sir Robert Armstrong, British cabinet secretary, in the 'Spycatcher' trial in 1986, who, attempting to prevent the publication of a book by a former MI5 employee, said '*It contains a misleading impression, not a lie. It was being economical with the truth.*'

Here is an example of how to spin by leaving out essential information and a tip on how to spot it from a post on my blog *Allo-*

porus | Ideas for healthy thinking entitled 'Never leave a number alone'.

> *Suppose a Minister announces with great fanfare a further $4 million in funding for schools. Not bad you might think. It would take the average Joe a couple of working lifetimes or a lottery win to get that kind of cash. In Australia, $4 million is enough for the salaries and overheads of roughly 40 teachers for one year. These teachers would be expected to look after 560 students, quite a few it would seem. Only there are 1.52 million high school students in Australia, making 560 a tiny proportion of those seeking erudition and selfies. So the 40 extra teachers recruited from the Minister's largess would teach 0.036% of the high school population for one year. After that, the Minister would need to make another announcement.*

The announcement was factual and large enough to impress most people who would never see $4 million. The context was carefully omitted.

Suppressing the negative news

Spin can be achieved by ensuring that some stories are never told or told to just a few. There are plenty of ways to do this, and it has been a tactic ever since there have been media outlets to reach the people.

Spin doctors suppress stories by buying off the outlets, pressuring the reporters, flooding the media with alternative narratives, staging press releases when everyone is in the pub, and never releasing negative news on a slow news day. Of course, the simplest and most effective is not releasing the facts.

Play the person

Rather than present or debate the facts, make the issue about the person, even if that person is the innocent messenger. Playing the man rather than the ball is slander—*the action or crime of making a false spoken statement damaging a person's reputation*—and defamation involving leaking misinformation about opponents. And it can get ugly.

A sample of the false and defamatory statements made by Donald Trump about Hilary Clinton during the 2016 election campaign shows just how ugly slander gets in the modern era: She has a serious chronic illness. She is sleeping all the time. She founded ISIS "literally" with President Obama. She wants to eliminate the Second Amendment. She started the birther movement.

Founded ISIS, always asleep, and started the birther movement... for goodness sake. And it helped get him elected. What is wrong with people?

Grandstand for effect

Make America great again. Oh my, let's do it. Let's shout and parade and sing and be great. Well, this, of course, assumes that the country was great in the first place. By no means a given. Many worldwide would argue more for 'war mongering bully', but, hey, it's grandstanding.

It's a favourite place of the narcissist. The 'look at me' then 'vote for me' style of leader that has become way too prevalent in recent times.

Don't try to spin; just be ignorant

We could inject the disinfectant to clean the insides. Enough said.

Lie and deny

If the standard methods are not working, or you can't be bothered with any sense of decency in your spin, then there is always the blatant lie.

Today, we call this fake news.

> *a form of news consisting of deliberate disinformation or hoaxes spread via traditional news media (print and broadcast) or online social media.*

Fake news is designed to damage by spreading lies. It can be nuanced, but this is the base of it—lies are told for personal or political gain. Donald Trump is a master at the big lie, and lies of all sizes defined the political career of Boris Johnson that took him to Downing Street, where he carried on lying to the people and the parliament.

The bigger the lie, the grander and more sensational, the better. Size gives it more traction, especially on social media, the opposite effect of what logic might suggest. This happens because social media is about retention on the platform—the longer eyeballs stay, the more ads they see—and people remain engaged longer when a hefty argument occurs. The more extreme, the better.

An online platform publishing a story with false content that attracts users benefits advertisers and improves ratings and access to online advertising revenue. So it's not just politics.

Claire Wardle of First Draft News made this list with seven types of fake news:

1. *satire or parody—no intention to cause harm but has the potential to fool*

2. *false connection—when headlines, visuals or captions don't support the content*

3. *misleading content—misleading use of information to frame an issue or an individual*

4. *false context—when genuine content is shared with false*

contextual information

5. *impostor content—when genuine sources are impersonated with false, made-up sources*

6. *manipulated content—when genuine information or imagery is manipulated to deceive, as with a doctored photo*

7. *fabricated content—new content is 100% false, designed to deceive and do harm*

Incredibly, societies of educated people with free will and a standard of living that would have made ancient kings green with envy allow enough of this nonsense to persist that it can be classified into seven categories. What is wrong with us?

But the critical point in all of this is the intent.

Spin, fake news, post-truth, the whole misinformation shebang **moves everything away from objectivity**. Recall that objectivity is where personal feelings or opinions do not influence our decision-making process **in considering and representing facts.** Spin is the exact opposite. It says to forget the facts; it's the outcome that matters. And it gets there by triggering our emotions.

Here are some of the common spin techniques used to such us all in:

- **Selective disclosure** is when a company or organisation releases only favourable information to the media while withholding any negative or unfavourable details to create a more positive public perception. Also known as "*being economical with the truth*".

- **Framing** presents information that emphasises certain aspects or angles to shape public opinion or influence how an issue is perceived. For example, presenting a corporate restructuring as a move towards "efficiency" rather than "layoffs."

- **Cherry-picking data** is where only the data or statistics that support a specific narrative are presented while ignoring or downplaying any contradictory data to make a product, policy, or person appear more successful or popular.

- **Euphemisms** are using mild or vague language to soften the impact of adverse events or actions. For example, using "rightsizing" instead of "layoffs" or "correctional facility" instead of "prison."

- **Misdirection** is a technique to shift the focus away from negative issues or scandals by diverting attention to unrelated positive news or achievements, aiming to minimise the impact of the negative press—look at this big, shiny bauble over here.

- **Astroturfing** is the creation of fake grassroots movements or fake social media accounts to give the illusion of widespread public support or opposition for a cause or product.

- **False comparisons** are misleading or irrelevant to make a product or service appear better than its competitors.

- **Bandwagon effect** creates an impression of widespread support by using statements like "Join the millions who already use our product" to encourage others to follow suit.

- **Expert endorsements** use the reputations of experts, celebrities, or influencers to gain credibility and create a positive association with a product or cause, even if the endorsers have no real expertise in the area.

- **Fearmongering** exaggerates potential risks or threats associated with a competitor's product or policy to create fear or uncertainty among the public and steer them towards the preferred alternative.

Is everything fake?

Lips perhaps, the Donald certainly, but everything? Here is what the late, great humourist and author Sir **Terry Pratchett** (1948—2015) said about the spread of fake news on the Internet in 1995, almost 30 years ago.

"Let's say I call myself the Institute for Something-or-other and I decide to promote a spurious treatise saying the Jews were entirely responsible for the second world war and the Holocaust didn't happen and it goes out there on the Internet and is available on the same terms as any piece of historical research which has undergone peer review and so on. There's a kind of parity of esteem of information on the net. It's all there: there's no way of finding out whether this stuff has any bottom to it or whether someone has just made it up".

And so it is today.

There is little to no policing of information online; as Pratchett said, anyone can put anything up on a website and claim authority. Anyone can spin the facts, be economical, or lie through their teeth. It is never that easy to know what is true and what isn't.

Fortunately, you now know a bit about scepticism. The sceptic goes with the premise of **not believing anything you read or see on the internet** until you verify it with reliable evidence. Armed with scepticism, most post-truths and fake news items are easy to spot as long as you are vigilant and have some evidence or common sense in your back pocket. Spin is more difficult because the PR people are experts at pushing our buttons. They readily play on our emotions rather than objectivity to run their agenda.

We are all going to get caught by spin once in a while.

Communications experts peddle post-truth and spin. They hide behind their spokesperson, the politicians and media personalities, including the new breed of internet influencers. Still, they can rarely recruit legitimate scientists and researchers into their storytelling unless companies hire those people in think tanks or as lobbyists.

A quick check on the affiliation of the researchers providing the evidence goes a long way towards the necessary validation. We should expect that most university and public agency researchers are objective. As a generality, this holds.

But what about when it doesn't?

What about when the scientists spin it?

As we saw in our explanation of the scientific method, science is supposed to be objective. It is meant to ignore values and follow an unbiased path in the design and execution of methods to generate evidence.

Academia also has a quality control system, the peer-review process, where, before publication, research is critiqued by expert peers who report to editors on the quality of the science. It's not flawless, but it works. It is reasonably safe to assume that research in these peer-reviewed publications should meet a minimum standard of scientific credibility, especially if you filter it through an evidence hierarchy.

Then there is the human element.

Scientists are driven by curiosity, a desire to know why, exploration of new places, finding answers, and being the first to find the answer, along with all the usual human frailties. We like to win.

Sometimes, our humanity can get in the way of objectivity. And sometimes, we give in to our biases.

> *"Scientists are a bunch of sceptics trying to find out the truth and prove each other wrong"*
>
> Alan Dore

When information and evidence are generated based on an assumed, usually unstated, preference for a particular outcome, policy or class of policies or outcomes, we call that **normative science.**

When scientific information is presented in words and phrases such as ecosystem health, biological integrity, and environmental degradation, it is normative science because the nouns are qualified with value judgements.

Philosophically, normative science is an accumulated body of provisional knowledge that '*seeks to discover good ways of achieving recognised aims, ends, goals, objectives, or purposes*'. Each phrase suggests a preference—good versus bad—representing a form of advocacy expected in policy, but regular science is supposed to be objective. The whole rationale for the scientific method is to discover cause and effect or to establish the likelihood of evidence being true irrespective of the impact. There is no need for the outcome to be somehow beneficial—just ask the researchers at the Los Alamos Laboratory working on the Manhattan Project during World War II.

A wide range of skills are needed to be good at science. An extensive formal training program is a requirement, and those who would be the researchers who design the experiments must progress to a minimum of a master's degree or equivalent experience. These days, most will have obtained a doctorate.

Scientists generate evidence through the scientific method. Then, they question each other's results and repeat the process to settle the matter, easily driven by ego and curiosity. They employ a range of processes, critical thinking and scientific reasoning skills to get the scientific method more or less correct; there is little bandwidth left for social skills or fashion sense.

Scientists are also fallible. They can do all the human things, including lying, cheating, and being economical with the truth.

All this means for a mindful sceptic is that not all science will be objective. Evidence is not automatically objective because it comes from a famous university or was written up by a professor. Even evidence that has passed peer review can be flawed.

Mindful sceptics are well aware of this human frailty. And because we are all human, it pays to use your mindful scepticism all the time.

Key points from this chapter

- Spin is a form of propaganda achieved through disingenuous and deceptive tactics to influence public opinion and emotions rather than presenting objective facts.

- The prevalence of spin, misinformation, and propaganda in modern public relations, politics, and communication is a problem.

- Some standard spin techniques include selective disclosure, framing, cherry-picking data, euphemisms, misdirection, astroturfing, false comparisons, expert endorsements, fearmongering, and grandstanding for effect.

- Fake news spreads deliberate disinformation or hoaxes through traditional and online social media.

- The sceptic has the skills to stay vigilant and verify information with reliable evidence before accepting it as truth.

- Even scientific research is not immune to spin with the potential for normative science, where bias or vested interests of researchers influence the outcomes of studies.

Chapter Five

Why mindful scepticism?

Where should you use scepticism?

The short answer is everywhere and all the time.

In your imaginary life on the savanna 14,000 years ago with a handful of your kin, each day was the same as the previous, and a few survival skills got you through to the next full moon. By the time you could throw a spear, you knew how and had absorbed all the knowledge of your tribe.

Today is different. Today, your life is complex.

When you were big enough to throw a spear, you were told you couldn't because it was dangerous and you had to go to school. A few years later, you could read, write, and do a bit of unnecessary algebra but still had so much to learn. Then you left school and realised that the well-meaning teachers hadn't taught you much real-world knowledge. Now, you had to learn new rules for the workplace, college, nightclub, the government, and the used car salesman. Soon, it was clear that street smarts helped a great deal but mostly applied to your neighbourhood. There was still so much to learn about everywhere else.

Fortunately, you now have help in this quest for practical knowledge.

You have access to endless information at the flick of a thumb. In your hand is a device capable of searching through much of the knowledge accumulated by humans over thousands of years and delivering a version of that knowledge almost instantly. Now you have a problem that your ancestors could not even dream about—too much information.

Even your focused searches deliver dozens of content options. By changing your attitude towards knowledge, you avoid the pain of being overwhelmed. The tool in your hand becomes a source of specific, uncontested details, such as the nearest wood-fired pizza outlet, directions, or an Uber to take you there. You use technology to simplify your complex life, which is now totally dependent on energy, resources, and other people.

Luckily, the device also has TikTok and Instagram. You escape into the phone to receive the stimuli that generate dopamine in your brain and feel much better. Dopamine is a neurotransmitter synthesised within certain neurons in the brain that acts as a chemical messenger between nerve cells (neurons) to trigger brain activity associated with several actions and functions. It plays a vital role in human behaviour, cognition, and emotions and interacts with various brain regions and systems to shape our experiences and responses to the world around us.

And now, instead of information, you get reward and pleasure. Only this situation is unsustainable. The complex world that you live in is fragile, vulnerable and prone to collapse. It has evolved to feed you stimuli and a protein-rich diet to produce dopamine because the triggers are profitable—eyeballs and clicks—but has ignored most consequences to you, your tribe, humanity and the planet.

You cannot survive on dopamine alone. You also need some of the stuff the six-continent supply chain provides and want a lot more. Before long, another delivery arrives.

Only fossil energy and resources are finite, as is the land's and oceans' capacity to absorb waste energy and materials. Meanwhile, the economic system that supports you assumes the opposite because it requires growth or it fails.

Humanity has itself a problem.

We break all this down in our *Mindful Sceptics Guides* that cast a mindfully sceptical eye over the critical issues from biodiversity loss to global food security and sustainable diet. But before that content takes us into the details, I want to show you how a mindful sceptic has a better time on social media and copes with the deluge of spin, fake news and modern politics.

She applies scepticism with mindfulness.

Why mindful scepticism?

Sceptics are expert questioners who learn how and when to ask the right questions. Blessed with boundless curiosity, they explore and can remain open-minded in the face of the most entrenched dogma. Because they want answers, they gain skills in logic, the scientific method and evaluation to find reliable evidence.

Sceptics are the torch-bearers of critical thinking, fact-checking, and evidence-based decision-making. Without sceptics, there would be an increase in misinformation, gullibility, and a lack of progress in knowledge and understanding. Because of what they do and what happens when they are sidelined, sceptics have contributed to every stage of human development, even when they are lone voices and the majority don't want to hear them.

Humanity wouldn't be here without sceptics, and the future of humanity depends on them.

Even if you don't want to be a lone voice, there are personal benefits to being a sceptic. You are better at coping with information overload and know all about dopamine. You can keep the bullshit artists at bay. Even though you know more than most about the planet's fragility and human socio-economic systems, your eco-anxiety is kept in check because you know the facts from the fiction, the truth from the spin.

Being a sceptic makes sense. **If you want to help yourself, your fellow humans and the planet, the advice is to become sceptical.** So why the mindfulness modifier?

It has to do with focus. The sceptic has curiosity and objectivity but not always awareness. It can be hard to know what to be sceptical about. Sceptics can't question everything because that is exhausting and impractical. Instead, they need a technique to focus deduction skills in the most valuable areas. And to do that, the sceptic has to become aware, appreciate the context and learn to ask better questions.

And a great way to achieve all three is to be mindful.

Curiosity doesn't kill the human

Curiosity is a natural human trait characterised by a strong desire to explore, learn, and seek knowledge about the world and its various phenomena. It is a powerful motivator that drives individuals to inquire, investigate, and discover new information or experiences. Curiosity sparks a sense of wonder and fascination, prompting people to ask questions, explore, and remain open-minded in their quest for understanding.

Sceptical humans have a boundless curiosity that displays as:

- Inquisitiveness—curious individuals are strongly inclined to ask questions and seek answers. They are not content with surface-level knowledge and are motivated to delve deeper into a subject.

- Openness to new experiences—curiosity makes people receptive to new ideas, perspectives, and opportunities. It encourages them to leave their comfort zones and embrace novel experiences.

- Critical thinking—curious individuals use critical thinking and analysis to make sense of information, solve problems, and seek evidence to support or refute beliefs.
- Creativity—curiosity often fuels creativity and innovation. It encourages individuals to think outside the box, explore alternative solutions, and connect seemingly unrelated concepts.
- Persistence—curiosity can sustain individuals in pursuing knowledge to overcome obstacles and setbacks.
- Lifelong learning—curious individuals love learning and actively seek opportunities to expand their knowledge and skills.
- Motivation—curiosity is intrinsic motivation, inspiring individuals to explore and discover without external rewards.

And who wouldn't want to have these characteristics? They are great because curiosity plays a significant role in human development, education, and scientific exploration. It leads to acquiring knowledge, personal growth, and a deeper understanding of the world and oneself. Cultivating curiosity can foster a sense of wonder, promote intellectual development, and enrich one's life by opening doors to new opportunities and experiences.

Mindfulness

Mindfulness is a mental practice for awareness that involves **paying deliberate attention to the present moment without judgment**. The idea is to be fully present and engaged in the here and now, observing one's thoughts, emotions, and sensations with curiosity and acceptance.

Most of the books and advice on mindfulness live in the self-help section of the bookstores because being present can help reduce stress, enhance focus, improve emotional regulation, and foster a greater sense of well-being.

It's a personal thing.

Mindfulness techniques are often used in meditation practices and have been integrated into various therapeutic approaches to promote mental clarity, self-awareness, and overall mental and emotional balance. Clarity, awareness and balance are a powerful trifecta to add to the sceptic's curiosity, critical thinking and objectivity. Some standard mindfulness practices include:

- Mindful breathing—involves paying close attention to the breath entering and leaving the body, which helps anchor the mind to the present moment and promote relaxation.

- Body Scanning—directs the attention to different parts of the body, noticing sensations and any tension or discomfort to enhance body awareness and help release physical tension.

- Mindful meditation—sitting quietly and observing thoughts and feelings as they arise without judgment to redirect attention to the present moment whenever the mind wanders.

- Loving-kindness meditation—cultivating compassion and kindness toward oneself and expanding it to loved ones, acquaintances, and even tricky individuals.

- Mindful eating—focusing on the sensory experience of eating, such as the taste, texture, and smell of food, to building a healthier relationship with food and make more conscious eating choices.

- Mindful movement—paying attention to bodily movements, breath, and sensations during yoga, tai chi, or walking meditation.

- Mindful listening—paying attention to the words, tone, and non-verbal cues without interrupting or forming judgments during conversations.

- Mindful appreciation—practising gratitude and appreciation for the simple things in life, especially nature, to enhance feelings of contentment and well-being.

- Mindful awareness in daily activities—engaging in routine activities mindfully, such as washing dishes, brushing teeth, or walking, to transform these moments into opportunities for presence and awareness.

A quick search on YouTube and dozens of practitioner channels will provide access to these techniques and more. Many of them are excellent and will show you that bringing these mindfulness practices into daily life is easy enough. They will also tell you that regular practice of mindfulness techniques can lead to increased self-awareness, reduced stress, improved focus, and enhanced overall well-being.

Alright, now the sceptic is horizontal, stress-free and wearing a kaftan. Why?

Once the sceptic is mindful, there is so much more to see.

Let's start with the people.

8 billion people

What do mindful sceptics say about 8 billion people increasing at over 8,000 an hour? What questions do they ask? Here is how we describe the context in the introduction to our *Mindful Sceptics Guide To The Human Escape From The Malthusian Trap*

> *The population spike that humanity is in right now is unprecedented in evolutionary history. There's never been a single species able to appropriate resources and energy to itself at such a rate and extent as* Homo

sapiens *have since we figured out how to use exogenous energy. In short, we eat and drink fossil fuels and turn them into more people.*

The neoliberal politician, his corporate buddies, and the talking head economists see population growth and rejoice—more votes and taxes, more sales, and economic theories are proven. Other people, blessed with an ounce or two of common sense, know that whilst more people might be good for business, sooner or later, the resources and energy that power the commerce people undertake will become scarce. They know this because the planet is finite.

They also know there is only so much waste that a finite planet can absorb. A population spike is both good and bad news.

Recall that the **sceptic** can question ideas and think objectively no matter what cultural, social or personal beliefs are in the mind. A **mindful sceptic** is balanced and assured of what is fact and fiction. She has clarity, awareness and balance to add to doubt, curiosity and objectivity. So what might a mindful sceptic think about the contentious issue of population, one topic we are not allowed to ask about?

Understanding the population relies on facts, mainly how many people there are now and how this compares to the past. It also helps to know some basic ecological theories and predictions for the future. Then there are values. More people are good for some, worrying for others, an opportunity and a cost, so it is riddled with values. There is also the fundamental value of reproduction that is innate to our biology and that some see as a human right.

All these attributes and concepts need some evidence.

Well, the numbers are verifiable. Almost all countries have reliable population censuses because governments need a handle on demography to plan for services and their tax base. Census methods are robust, and other ways exist to verify the data. Historical estimates can become sketchy, and pre-history estimates of the human population are subject to more error, but the broad trend in the population data is hard to miss. The sceptic can be confident in the human population numbers and the ecological theory, which predicts that when a plant or animal species has access to unlimited resources and is not checked by predators or disease, numbers will increase exponentially at a rate determined by the species' reproductive rate. Ecologists have described this prediction's logic, mathematics, and mechanisms, observed it in nature, and replicated it in the laboratory—exponential population growth under unlimited resources is settled science.

A diligent sceptic might want to review this evidence in the textbooks or the primary literature but would realise if the evidence is strong and no longer contested, then exponential population growth is a thing if the conditions under which it occurs are real.

Solid textbooks on population ecology

- Begon, M., Mortimer, M., & Thompson, D. J. (2009). *Population ecology: a unified study of animals and plants.* John Wiley & Sons.
- Begon, M., & Townsend, C. R. (2021). *Ecology: from individuals to ecosystems.* John Wiley & Sons.
- Vandermeer, J. H., & Goldberg, D. E. (2013). *Population ecology: first principles*. Princeton University Press.
- Roughgarden, J. (1979). *Theory of population genetics and evolutionary ecology: an introduction*.

Accepting that exponential population growth is possible, even a casual review of the data for the global population of humans shows that the trend follows an exponential path—a slow change that becomes rapid as numbers increase.

Here are the numbers on a standard linear graph.

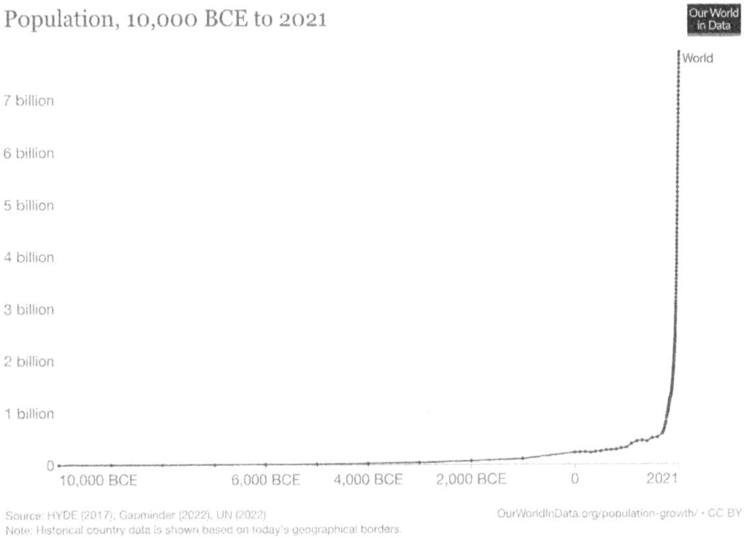

Ecological theory is explicit in that exponential growth can only happen with unlimited resources—for animals this is food, water, and space—that allows most individuals in a population to survive, grow, and reproduce. Assuming there are no other constraints like disease or hungry lions, bears or wolves.

We felt hungry when we imagined ourselves on the African savanna 14,000 years ago. The main agenda for the day was finding enough food. Along with the risk from the lions, cobras and malaria, there was not enough food for more than a handful of humans to survive, grow and reproduce in any one place.

The sceptic knows this quickly enough. Indeed, blind Freddie knows. How long would a tourist on a safari to Africa last if they had to find food outside the luxury tented camp with nothing but a rudimentary spear, a flint and a gourd to hold water? A few people might make it but couldn't beget a dynasty of millions. There is not enough food.

So how come? What explains the numbers?

Freddie chimes in with the obvious answer of agriculture. Humans figured out how to grow food. Fair enough, but we are sceptics. Remember, we need evidence of the connection between agriculture and population. We unpack the history of food production in our *Mindful Sceptics Guide to Growing Enough Food,* and the sceptic would dive into that guide and follow up on the evidence it describes. Assuming you have done that, you now know that, for a long time, agriculture produced enough food for a significant growth in human numbers at an exponential rate but with a small exponent delivering an annual growth rate of 0.04%.

People with agriculture had more food than hunter-gatherers, sufficient to stay put and build complex societies, even great cathedrals, cities and viaducts. Still, there was a limit to how much food could be grown. The great cultures dragged food in from far and wide and relied on most of their people and as many enslaved people as they could oppress to put their energy into growing the food. Agriculture was a net energy exporter, but only just. Population growth was relatively slow because agriculture was inefficient, and there were disease and nutrition constraints on survival and reproduction.

Agriculture does not explain the spike.

The sceptic takes a closer look.

If the recorded growth rate through our early agricultural history was 0.04% and the global population was 4 million in 10,000

BCE, then by the year 2000, there should have been roughly 490 million humans.

If there were more hunter-gatherers at the beginning of agriculture, say 8 million, double the current best estimate, there would be 970 million people by 2000. The evidence is that over 6 billion people were alive in 2000, **six times more** than the generous estimate on a 0.04% growth rate.

So let's go to a more reliable start number, say the 2.5 billion people estimated to be on the planet in 1950 when all regions of the world were known, and each country's population census was robust. Then let's calculate the growth in numbers at 0.04% for 71 years—we get 2.57 billion, just 72 million more people. In reality, we added 5.41 billion more people, 75 times the number at a 0.04% growth rate.

Clearly, the population grew faster than 0.04%. In fact, to increase from 2.5 billion in 1950 to 7.9 billion in 2021 requires a population growth rate of 1.75%.

The sceptic concludes that something must have changed. She will ask why and look for some causes for the effect. The details are explored in the *Mindful Sceptics Guide to Escaping the Malthusian Trap,* but the short answer is that humans produced more food thanks to the energy from fossil fuels and kept more people alive with technology.

A sceptic is fond of these back-of-envelope calculations—maths is your friend—because they are excellent logic tests and make it easy to know which numbers must be verified. A mindful sceptic is brave enough to tackle the taboo topic of population because she has the skills of curiosity and logic to review the evidence and, from the mindfulness, the clarity, awareness, and balance not to be put off the investigation.

When a mindful sceptic digs deeper into the population issue to figure out the basics and then maybe learn from experts—Hans Roslings *Factfulness* is an excellent place to start—then the population becomes what it is for the future, a challenge at least and a potential disaster. She asks if 8 billion people can find enough food.

Global food security

Eating is terrific; everyone likes to do it. And for many, food is delightful, for there is magic in cooking and preparing the ingredients that nourish your loved ones.

Then, there is the reality that food can also be a pain. It costs money, has to be carted from the store, takes time to prepare and makes a mess. It even takes some planning, or the worst happens, and there is no milk for the Cheerios.

If we asked a polite Jane as she pushed her trolley around a Western supermarket what mattered to her about food, here is what she might say: keep food cheap; make food easy to buy, prepare and eat; make sure food can be eaten while holding a phone; the kids must want to eat it; never run out of banana Twinkies.

But most people don't shop with a trolley. The World Bank estimates that nearly half the world's population—over 3.5 billion people—still struggle to meet basic needs. On average, these people survive on less than $5.50 per day to pay for everything that comes with a bill. Often, close to half this amount is spent on food.

Could you manage to feed yourself on $2.50 a day for food? At a push, you might just survive on instant noodles. The billions of people who can't are struggling with food security.

A standard definition of food security is **that all people have access to sufficient, safe, nutritious food at all times to maintain a healthy and active life**. This utopian condition for humanity is the commonly accepted definition of food security developed during the 1996 World Food Summit and widely adopted by the World Bank and the UN Sustainable Development Goals. A shorter option to the long-winded definitions is **feeding everyone well.**

There is no ambiguity about everyone all the time, sufficient, safe, nutritious, health or well-being.

Each word and phrase is clear and combines into the most egalitarian and humanitarian definition in the global Lexicon. Whilst politics is ever present, it is hard for any political party or politician to sour an ambition to feed everyone well.

Of course, Jane may be a good cook, worry about healthy eating and have an eye on food miles because she knows that a long supply chain is terrible for the environment. She may even learn a little about nutrients, the inflammatory consequences of too much sugar and grains, and the need to avoid ultra-processed foods. And she values the delight in good food, especially when shared with friends and family.

Given how we guess Jane feels about food, we might add to her list of concerns: keep fresh fruit and vegetables available; ensure low-salt, low-fat, and low-calorie options on the shelves; make reusable bags compulsory; reduce packaging; and pin the best recipes on Pinterest.

Despite the vast number of products, the supermarket might only meet some of these needs. Jane may also buy food from local growers, choose organic options from local suppliers, and perhaps cook vegetarian meals for her family.

Even if Jane is a bit diet conscious and worried about the environment, it is doubtful that she will know why the global food supply is insecure, what must happen to feed everyone well, or that food prices have gone down proportionately in the last 50 years.

But if Jane were a mindful sceptic, she might bring to mind some of the items on the following list that are generally outside the average consumer's mind:

- overreliance on intensive agriculture (the energy subsidy)
- soil degradation
- soil biodiversity loss
- fertiliser inputs that come from fossil fuels
- long 'just-in-time' supply chains
- overreliance on food imports

- single source supply
- climate change impacts on production volume and efficiency
- micronutrient deficiency in foods
- carbon emissions from agriculture.

Because she is skeptical about what the advertising tells her and what the media keeps saying about eggs, she will change her mind about some items on her shopping list.

Values and decisions—the non-scientist sceptic

As a scientist in and out of the academic system throughout my career, I have learnt that science is a particular lens through which to view the world. It can be like looking down a dark tunnel and only seeing part of the vista as a dim light at the end. But once in the tunnel, most of us scramble towards the light by focusing closer on the narrow part of the world we can see and employing more specialised techniques and jargon to get there.

Add the human element to this narrow scientific focus, especially the competitive ego that is just as brittle in the ivory towers as it is in the boardroom and, well, objectivity can be lost.

The sceptic need not be so blinkered, and a mindful sceptic has the awareness to be much more pragmatic. She knows the human world is value-laden, replete with bias and delusion, and prone to self-interest. But she also knows that humans are intelligent, ingenious and able to cooperate. We also have imagination, creativity and artistic talents that add a secret sauce to our present and future.

In other words, you don't need to be a scientist or be able to conjure up and interpret back-of-envelope calculations to be a mindful sceptic. Nor do you need to be a philosopher with the skill to disappear into the rear end of language to obscure your thoughts from the masses.

All you need to do is be curious and aware and apply critical thinking.

Everyone can do it.

Life without sceptics

If you have followed the description of a mindful sceptic this far, you will know the what and why of these valuable humans.

Rather than go over all the reasons why humanity has to have sceptics, why it must support them, build them up and listen carefully to their questions and demands for reliable answers, let's conclude this chapter by imagining what would happen without scepticism.

As the discussion of the misuse of evidence shows, we are heading into this version of society where people are susceptible to accepting information, beliefs, and claims at face value without questioning their validity. Where dogma is dominant, and people outside your tribe are the enemy.

What else happens if humans lack critical thinking? Here are a few possible consequences of a lack of scepticism:

- Vulnerability to misinformation—people become more likely to accept false or misleading information, whether intentional misinformation or unintentional errors, leading to spreading rumours, conspiracy theories, and baseless beliefs.

- Increased gullibility—where people tend to trust everything they hear or read without verifying the facts, leading to exploitation by those with ulterior motives, such as scam artists or dishonest advertisers.

- Stagnation of knowledge—without a questioning attitude, scientific discoveries and advancements are hindered, along with the pursuit of truth.

- Blind acceptance of authority—because scepticism encourages individuals to question authority and demand evidence (a big reason why its frowned upon), without it, people might blindly accept the statements of leaders and authority figures without holding them accountable.

- Diminished critical thinking—underdeveloped abilities to analyse information objectively and make reasoned judgments

- Impaired decision-making—the tendency to make choices based on flawed or incomplete information, leading to poor outcomes.

- Erosion of trust—people can feel deceived or manipulated due to their uncritical acceptance of information.

Without scepticism, there is less critical thinking, fact-checking, and evidence-based decision-making, along with increased misinformation, gullibility, and a lack of progress in knowledge and understanding. None of this is good when humanity faces many acute, complex and wicked challenges and needs all the brainpower to find and deliver imaginative solutions.

Assuming you are now convinced that mindful sceptics are useful people to have around, we should look closer at how to become one.

Key points from this chapter

- Scepticism should be applied everywhere and always, especially in today's complex and information-rich world, to help discern reliable evidence from misinformation and propaganda.

- An abundance of information can lead to information overload, especially when we rely on technology to simplify our lives, seeking immediate rewards and pleasures.

- A mindful sceptic combines curiosity, objectivity, and awareness to ask critical questions and focus on crucial issues that require evidence-based decision-making. By being mindful, a sceptic avoids normative biases and remains open-minded.

- Even sceptics can be fallible and may succumb to biases or misinformation. It is essential to approach information critically, verify evidence, and question assumptions to foster a deeper understanding of complex issues.

- The future of humanity depends on sceptics challenging the status quo, promoting critical thinking, and driving progress in knowledge and understanding.

Chapter Six

How to be a mindful sceptic

Rules of thumb for the mindful sceptic

Before we dive into the details of the many skills and techniques for being a mindful sceptic, which might be a bit much, here is a simple rule of thumb used by the mindful sceptic during the daily rounds

Stop—Think—Listen—Act

Stop scrolling
Stop listening to gossip
Stop to quiet the mind and breathe

Think for a moment
Think about context
Think about what could be fact or fiction

Listen to yourself
Listen to others
Listen to the wind

Act by consuming information
Act by evaluating evidence
Act as an advocate for scepticism

Being a mindful sceptic

Being a mindful sceptic is a choice. It starts with a mindset that juggles curiosity and doubt; for most of us, that is enough.

Being thoughtful about the information and opinions that come our way can reduce information overload and dismiss the worst of the misinformation and bullshit designed to annoy us and relieve us of our cash.

Once the mindset changes, then curiosity grows. Instead of doubting the information, a mindful sceptic embraces intellectual curiosity and wants to know what is happening. This is where the questioning starts. The what, how, where and why questions flow, and the search for answers begins.

The sceptic side seeks knowledge from the vast repositories of evidence that humans have accumulated and curated over the centuries and from the application of science, which is our most effective method for finding the facts. Knowledge gathering can mean new skills and time spent going to the source and understanding the evidence process—more on this in a moment.

The mindful side recognises that even the most reliable evidence only becomes knowledge when humans use it and that humans have biases, beliefs and hold values close. We are only sometimes logical, sensible or just. Mindfulness helps to find the context for the evidence and the balance between what the evidence suggests and the messy reality of human systems.

When these two combine, a mindful sceptic has access to awareness, embraces uncertainty and avoids the pitfalls of the cynic and the gullible.

A mindful sceptic strives to be a well-informed and discerning consumer of information and a source of ideas for solutions to any number of problems.

If you prefer a listicle, a step-by-step guide on how to cultivate mindful scepticism, here is roughly how the sequence goes:

1. **Embrace intellectual curiosity**—foster a genuine de-

sire to learn and understand more about the world by engaging with diverse topics, seeking out reliable sources of information, and remaining curious about different perspectives and viewpoints.

2. **Question and scrutinise**—develop the habit of questioning ideas and beliefs, including your own, by asking for clarity in definitions, consistency in logic, and adequacy of evidence. Be willing to engage in thoughtful inquiry and fact-check claims before accepting them as genuine.

3. **Be aware of cognitive biases**—recognise that all humans have biases that can influence beliefs and decisions. Stay mindful of emotional influences and try to stay objective when evaluating information.

4. **Consider ethical implications**—reflect on the potential impact of your beliefs on yourself and others. It's a value judgement, but consider promoting ideas that lead to positive outcomes for individuals and society.

5. **Engage in constructive dialogue**—be willing to listen to different perspectives and engage in respectful discussions with others. Use these interactions to challenge your beliefs and refine your understanding of the world.

6. **Recognize the value of expertise**—be open to various viewpoints but acknowledge that not all opinions carry equal weight. A rapper's opinion on climate is not more

valuable than yours because they have 10 million views on TikTok. A highly trained and experienced climate scientist might be more reliable. Seek experts and consider their knowledge and expertise when making informed decisions.

7. **Embrace complexity and uncertainty**—because life is filled with ambiguity and uncertainty, avoid seeking simplistic answers to complex issues—Friar Ockham notwithstanding—and be comfortable with the idea that some beliefs may have varying degrees of certainty.

8. **Find balance**—look for the middle ground between scepticism and open-mindedness to avoid falling into the traps of cynicism or gullibility. Be receptive to new evidence and ideas while remaining critical and discerning about the reliability of information.

9. **Refine and adjust beliefs**—be willing to revise your beliefs based on new evidence and thoughtful reflection because being a mindful sceptic is ongoing.

10. **Apply to everyday life**—practice mindful scepticism from personal interactions to engaging with news and information on social media. Strive to be a well-informed and discerning consumer of information.

By incorporating some or all of these principles into your thinking and decision-making processes, you can become a mindful sceptic who navigates the complexities of the modern world with grace, wisdom, and a commitment to seeking truth and understanding.

Scepticism is about doubt—and curiosity

At its core, scepticism is about doubting almost everything.

The sceptic is primed to question and unlikely to take many things at face value. Before accepting a premise or forming an opinion, the sceptic must understand what it is and where it came from and evaluate the evidence for the statement or the situation.

Philosophers have grappled with this challenge for centuries. What is knowable or what is not knowable? And whether or not you can live with a balance between the two or accept certain things as being true but unknowable. Even our brief history in chapter 2 shows how convoluted this can become.

It is hard to learn how to doubt, but doubt is at the core of scepticism.

The definition of disbelief—not believing anything until you are satisfied that your level of rigour in the evidence has been reached and that you are confident in the available evidence—can help. What you do as a doubter is to be uncertain about a belief, claim, or proposition, and that involves questioning the truth, validity, or reliability of something that arises when there is insufficient evidence or conflicting information. Doubt should be a natural and healthy response when faced with new information, complex situations, or conflicting opinions, as it prompts us to seek further understanding or evidence before accepting or rejecting an idea.

While it can be a positive and constructive force, excessive or persistent doubt can lead to indecisiveness or hesitation. Striking a balance between critical thinking and an open-minded approach to new information is essential in navigating the complexities of doubt. Gathering additional evidence, seeking expert opinions, or engaging in open and respectful discussions to better understand the subject at hand will help. How you do all this depends on preference, aptitude and the issue's importance. And as I did with Keneilwe, you can doubt but choose to act as though all he said was true.

Not just doubt

Scepticism is also about curiosity. A feeling of wanting to know the truth neatly balances the negative emotions from doubt. There is positivity in wanting to find the truth and to learn as much evidence as possible. And there is feel-good in seeking out evidence.

If I see a video on TikTok where some metro dude is telling me that I need to do a specific breathing exercise to enhance my well-being, should I believe him? I don't know this guy from Adam and suspect he has zero qualifications in pranayama or any ancient art of breath control.

But I do know that those arts exist. And that controlling one's breath is a powerful way of calming emotions. So, I won't necessarily agree with everything he says straight away. Still, I'll know that he's coming from a tradition that has been studied and practised for hundreds of years for emotional benefit.

My curiosity would be whether I am as interested in understanding this particular breath technique and if it might be of assistance to me. I might take him on, watch the video, and think about it in a slightly more positive light than if he was selling me a drug to lift my mood.

When to apply scepticism is critical

The best tool for knowing when to be sceptical is mindfulness. The ability to pick the evidence, to think of the context in which the information is being presented, and the relevance of the context to your own life and well-being.

It matches those three things together to decide whether or not you need to be sceptical about this information or just let it slide.

So much of life we take for granted, but we also take it literally. When things have happened to us many times, we tend to assume that they're correct, but they may not be, and the information may not be. It's surprising how often scepticism is practical.

Apply scepticism when needed

It is easy for scepticism to be "*oh no, that's wrong*" and become the vehicle for being against everything. Conflation to contrarianism is how scepticism is hijacked by political extremes and the populism we see worldwide today. When sceptical becomes denial, so we can't use it in that way. We must use it in a healthy, positive way. And to do that means to know when the issue or the situation is relevant to you.

There are two areas where this is important.

One is when a critical moral issue is affecting you. This is easy to pick because the claims or argument produces an emotional response. So, if you feel angry, upset, annoyed, or frustrated by a particular statement or opinion presented, then it's a good idea to fire up your scepticism and examine the evidence.

Use scepticism when emotions are triggered.

The second time to use it is when the feelings may not be engaged, but there's a commonality or repetition to something that has become conventional wisdom. When in fact, there's no evidence for that conventional wisdom. And this can be for large areas of our lives like, for example, the foundational economics of our systems, which we talk about a lot in the *Mindful Sceptics Guides*.

Another acute example is the food we're expected to eat. Diet is one of these areas where conventional wisdom is everywhere, surrounded by equivocal evidence—the reality is that we don't know what we should be eating because it's always a unique combination for each individual. What we do know is that many of the foods we are told are healthy are not healthy at all. We should be very sceptical about diet recommendations and fad diets. Use that scepticism to land on something ideal for you as an individual.

All this brings us to one of the significant challenges facing the sceptic—finding and evaluating evidence. Time and skills are needed to locate and unpack evidence, and *Mindful Sceptics Expert Guide to Understanding Evidence* covers the details.

But for now, it's sufficient to be aware of two things: 1) your life will benefit from a sceptical lens because many falsehoods creep into the every day, and 2) you don't always have to ask those questions because you can let many things slide.

So the overall advice about when to use scepticism is **when knowing the truth is essential to you,** and that is when the issue at hand is meaningful in your life, or you feel as though you need to understand it fully.

A mindful sceptic is not just someone who argues for the sake of it.

The mindfulness modifier

The takeaway from all this is to **use scepticism wisely**.

Scepticism can and should be the paradigm of first choice. *I don't believe that* is the go-to thought, but you don't have to speak it. Better to have a filter that says, *well, I don't need to worry about whether this particular information is true or not. It won't affect me, and I can let it pass by.*

Recall that the essence of enlightenment is the ability to let things pass through without touching the sides and allowing the poisonous energy of the world to pass you by—until you can't.

We must confront with curiosity and critical thinking because if we don't, chaos is likely: personal and collective chaos.

We know that 8 billion people, increasing at 8,000+ an hour, all seeking betterment and more making, is pushing the planet's ecological limits. Humanity is stomping over the earth in size 12 hobnail boots. It's a problem. The fossil energy pulse and our innate ingenuity have given humanity a sustainability predicament. We are decoupled from the energetic constraints of our basic ecology, which has consequences.

We cannot let that slide.

Skills of the best sceptics

The best sceptics, from ancient philosophers to modern scientists and critical thinkers, all possess a strong foundation in critical thinking, research, and logical reasoning. They remain open-minded yet grounded in evidence, communicating effectively and encouraging others to engage in thoughtful inquiry, and they excel at critical thought, logic, evaluation, communication, and science literacy.

It's a substantial skill set.

Most of all, they are curious about the world. They are good at asking the right questions—and knowing when to ask them. I want you to be a great sceptic, perhaps to become famous. But you might need to brush up on some content and skills. The *Mindful Sceptics Guides* can help as they cover a host of crucial real-world issues, from biodiversity loss to sustainable diet, and our *Mindful Sceptics Expert Guides* help to understand the evidence and how to feed everyone well.

In the meantime, here are a few practical tips for a mindful sceptic that have enormous benefits for your sanity and don't need a doctorate.

In no particular order...

- **Check facts and sources**—practice fact-checking and verifying information using reputable sources and common sense. Develop the habit of assessing the credibility and reliability of the sources you encounter daily, especially on social media and news platforms.

- **Journal and reflect**—record your thoughts, beliefs, and reflections and review your entries to identify patterns of thinking, biases, and areas where you can apply more critical thinking.

- **Engage with diverse perspectives**—seek diverse viewpoints and engage in respectful discussions with people with different beliefs. This will help you gain a broader understanding of various issues and challenge your assumptions.

- **Critical thinking exercises**—complete exercises and puzzles that challenge your reasoning abilities. Try to mentally solve problems that require analytical thinking and evidence-based decision-making, such as how to feed everyone well.

- **Don't make assumptions**—this is one of The Four Agreementsthat Don Miguel Ruiz explains in his practical guide to personal freedom. Don't make assumptions about others, but also ask yourself why you hold certain beliefs and whether sufficient evidence supports them.

- **Practice intellectual humility**—be willing to admit when you are uncertain or wrong about a topic because no one has all the answers.

- **Stay informed on cognitive biases**—educate yourself about common cognitive biases and fallacies, especially how they can influence your thinking, then avoid them.

- **Seek feedback**—welcome constructive criticism and consider alternative viewpoints to refine your understanding of various subjects by encouraging feedback from trusted friends, colleagues, or mentors.

- **Attend workshops and listen to podcasts**—pay attention to others talking about critical thinking, scepticism, and scientific inquiry. We recommend The Great Simplification, Planet Critical, and Doonbergto get you started.

- **Cultivate a curiosity habit**—embrace a daily habit of curiosity by setting aside time to explore new topics, read diverse materials, and engage in lifelong learning. More podcasts and documentaries.

- **Practice mindfulness and meditation**—practices to enhance self-awareness and clarity of thought can help you recognise and manage emotional biases while staying present in the moment.

- **Educate others**—share your knowledge and skills in critical thinking with others by encouraging friends and family to be mindful sceptics, promoting a culture of intellectual curiosity and open-mindedness.

These are just a few practical methods you can add to daily life to enhance your ability to think critically and make well-informed decisions. Over time, these practices will become habits contributing to a more intellectually humble, discerning, and empathetic approach to understanding the world.

The core attribute of the best mindful sceptics

We don't know everything and never can. There is too much to know. After over 40 years of scientific thinking, I come across new information daily and regularly stumble into new areas of evidence.

Back in the late 1980s, I was a young lecturer at the University of Botswana and keen to get ahead. I heard that two experienced researchers from Sweden were in the country to continue their work on how browsing herbivores influence plants and vegetation. Its a big thing in Sweden because of all the moose, so why not look at the African herbivores?

"We'd be happy to meet with you," said Professor Christina Skarpe in an email, "we are staying at Mokolodi." *Where*, I thought, *never heard of it*. It turned out that Mokolodi Nature Reserve was just down the road from the university and was an ideal location for all sorts of education and research. The introduction to my backyard from Scandinavian visitors was embarrassing, but we met and collaborated on several research projects. Once I knew it was there, I took my students to Mokolodi for practical classes and used it as a study site in several research projects. A little humility allowed for a huge benefit.

Mindful sceptics apply humility to their intellectual pursuits.

Intellectual humility recognises the limits of one's knowledge, the willingness to acknowledge gaps in understanding and the openness to being wrong or learning from others. It involves a sincere and genuine desire to consider alternative perspectives, admit when one lacks knowledge on a particular topic, and refrain from overestimating one's expertise.

People with intellectual humility are open to the possibility that their beliefs, opinions, or conclusions might need revision in the face of new evidence or information. They value continuous learning and growth, recognising that knowledge is not static and that there's always more to discover.

In a world that often promotes assertiveness and a "know-it-all" attitude, intellectual humility is valuable for constructive dialogue, collaboration, and a deeper understanding of complex issues.

A little intellectual humility helps a mindful sceptic avoid cognitive bias and the echo chambers of groupthink. It's a critical skill.

Why you need to be a mindful sceptic

You will gain insights, get to know yourself and gain protection from the fearsome modern world.

Curiosity and critical thinking are dopamine enhancers. All that brain activity will make you feel great.

In a blink, you will be so much more aware and knowledgable; everyone will want to talk to you for your take on the world will be interesting.

Then there is the real reason.

Humanity needs you.

Now, it is true that not everyone can or should become a moral philosopher, research scientist, or think tank analyst, for there are only so many places in these professions, thankfully. We wouldn't get very far if everyone were drinking wine, wearing a toga and pontificating in Latin or disappearing down a rabbit hole of specialism.

No, humanity needs pragmatic answers to smart, actionable questions—dare I say, mindful scepticism. And we need this from everyone: you, me and a host of our buddies.

I have a hunch that humanity will survive because enough people question the dogma to tip us over into more suitable alternatives.

So, how can you become a mindful sceptic?

Be a mindful sceptic today

What to do now is become a mindful sceptic today.

All it takes is a mindset of curiosity and doubt. As Don Miguel Ruiz says, do not make assumptions. Instead, seek the facts of the matter and listen to the evidence, not the messenger's opinion.

You don't need a research degree or a background in science, but it helps to have some content knowledge. It is hard to ask the right questions without the context in which to ask them.

The first thing to do is **increase your awareness of humanity's global predicament**—8 billion people living in an economic techno-system pegged to fossil energy that generates inequality and degrades the environment—and the consequences of a six-continent global supply chain to support the people and their pets.

Chances are you know something of this already, and we have our *Mindful Sceptics Guides* on topics from biodiversity loss to sustainable diet and global food security. These guides cover the issues through the mindful sceptic lens of objectivity and understanding evidence.

Our *Mindful Sceptics Skills* short course builds on this chapter with the 'how to' of critical thinking, questioning, evaluating evidence and making decisions.

And if that seems too easy, then there is *Mindful Sceptics Expert Guide to Understanding Evidence,* which takes a much deeper dive into the skills of the scientific method or *Mindful Sceptics Expert*

Guide to Feeding Everyone Well, which goes into the details of how we can make food, ecology and diet come together to deliver food security to 8 billion humans increasing at 8,000 an hour.

More education will certainly help, but the truth is you already have the heart of a sceptic and a mindful soul; we all do. It is the human condition. Only that innate ability for curiosity and critical thought is too often suppressed, beaten back by the pressures of modern life.

Perhaps all you have to do is take a breath and give yourself some space.

Reminder of the Rules of Thumb

Stop—Think—Listen—Act

Stop scrolling
Stop listening to gossip
Stop to quiet the mind and breathe

Think for a moment
Think about context
Think about what could be fact or fiction

Listen to yourself
Listen to others
Listen to the wind

Act by consuming information
Act by evaluating evidence
Act as an advocate for scepticism

Key points from this chapter

- Being a mindful sceptic involves combining curiosity and doubt to approach information thoughtfully and critically.
- A mindful sceptic questions and scrutinises ideas, seeking

evidence from reliable sources and being aware of cognitive biases.

- Sceptics seek knowledge from evidence and science, while the mindful modifier allows the sceptic to consider the human context.

- There is a balance between scepticism and open-mindedness that helps to avoid cynicism or gullibility.

- Practical tips and skills for being a mindful sceptic include fact-checking, critical thinking exercises, and seeking diverse perspectives.

- Intellectual humility and mindfulness help to manage emotional biases and enhance self-awareness.

- The goal of a mindful sceptic is to become a well-informed and discerning consumer of information, applying scepticism wisely for a more enlightened approach to understanding the world.

About the Author

Mark is an ecologist, not a green one. He has built, run and sold companies, held academic posts at four universities, advised governments on environmental matters, has a PhD and still plays too much golf. His ecological research ended up in more than 70 peer-reviewed research papers and over 80 technical reports, and he supplied peer-review for more than a dozen international science journals. He lived in Africa for a decade, where he researched termites and millipedes, and once walked into a lion who snarled at him. Much of his career has been in Australia, where he won an Outstanding Teacher award from Macquarie University. His consultancy practice advises commercial and public sector clients on assessing, monitoring, and evaluating environmental impacts, especially in carbon asset and liability management. His non-fiction books are about awareness of humans' role in nature and, as Paul Sorol, he chronicles and writes fiction for a change.

Click or scan for link in bio

Mindful Sceptics

MINDFUL SCEPTIC GUIDES

Abatement

Biodiversity loss

Escaping the Malthusian trap

Farmers

Food security

Food, people and soil

Growing enough food

Healthy and sustainable diet

Sustainability

The soil carbon debt

mindfulsceptics.info

Made in the USA
Monee, IL
04 November 2026

67379692R00079